Nonparametric Methods in Statistics with SAS Applications

CHAPMAN & HALL/CRC
Texts in Statistical Science Series

Series Editors
Francesca Dominici, *Harvard School of Public Health, USA*
Julian J. Faraway, *University of Bath, UK*
Martin Tanner, *Northwestern University, USA*
Jim Zidek, *University of British Columbia, Canada*

Texts in Statistical Science

Nonparametric Methods in Statistics with SAS Applications

Olga Korosteleva

California State University, Long Beach

CRC Press
Taylor & Francis Group
Boca Raton London New York

CRC Press is an imprint of the
Taylor & Francis Group an **informa** business

A CHAPMAN & HALL BOOK

CRC Press
Taylor & Francis Group
6000 Broken Sound Parkway NW, Suite 300
Boca Raton, FL 33487-2742

© 2014 by Taylor & Francis Group, LLC
CRC Press is an imprint of Taylor & Francis Group, an Informa business

No claim to original U.S. Government works

Printed on acid-free paper
Version Date: 20130715

International Standard Book Number-13: 978-1-4665-8062-6 (Paperback)

Visit the Taylor & Francis Web site at
http://www.taylorandfrancis.com

and the CRC Press Web site at
http://www.crcpress.com

Preface

This book has been written as a textbook for the second-year graduate course taught by the author in the Master's program in Applied Statistics at California State University, Long Beach. The goal of this course is to teach applications of nonparametric methods in statistics, starting with the tests of hypotheses, and moving on to regression modeling, time-to-event analysis, density estimation and resampling methods.

Being a textbook, this book has abundant examples and exercises. The settings were taken from various scientific disciplines: health sciences, psychology, social sciences, education, and clinical trials, to name a few. The settings and properly disguised data came from consulting projects that the author has been involved in over the past decade.

All examples and exercises require the use of SAS 9.3 software. In the text, complete SAS codes are given for all examples. To prevent typing errors, the large data sets for exercises are available at http://csulb.edu/~okoroste/nonparam.html. Qualified instructors may request the solutions manual for all exercises on the publisher's book companion website.

The author would like to thank the anonymous reviewers for providing constructive comments and suggestions, and also wishes to thank the editorial and production teams at CRC Press for all their hard work, especially David Grubbs, Marsha Pronin, Amy Rodriguez, and Shashi Kumar.

The Author
September, 2013

Contents

Chapter 1

Hypotheses Testing for Two Samples

Generally speaking, statistical methodology is split into two broad classes: parametric and nonparametric. *Parametric* methods assume that forms of distributions of random variables are known and only a fixed number of parameters of these distributions must be estimated. In many applications of parametric methods, even if the exact form of the distribution is not known, observations come from a sufficiently large sample, thus the use of the normal approximation is validated by the Central Limit Theorem. In the situations when the sample size is relatively small and no conjecture may be made regarding the functional form of the underlying distribution, *nonparametric* methods are used as an alternative.

In this chapter, we present the most widely used nonparametric hypotheses tests for matched paired samples and for two independent samples.

1.1 Sign Test for Location Parameter for Matched Paired Samples

If the matched paired observations are normally distributed or may be approximated by normal distribution, the paired *t*-test is the best statistical test to implement. If, however, the data are obtained from a non-normal distribution or one containing outliers, a nonparametric *sign test* is a better option.

1.1.1 Testing Procedure

Suppose we have n pairs of observations of the form $(x_i, y_i), i = 1, \ldots, n$. We assume that each pair is recorded either for the same individual, or for two different individuals who were matched with respect to certain characteristics on which confounding is not desired. The research question in mind is whether the observed variables come from the same distribution or the distributions differ in location parameters. Denote by θ_X and θ_Y the *location parameters* of the unknown cumulative distribution functions $F_X(x)$ and $F_Y(y)$, respectively. Typically, the location parameter is the mean, median or mode. The tested hypotheses may be stated as $H_0 : \theta_X = \theta_Y$ against $H_1 : \theta_X > \theta_Y$, or $H_1 : \theta_X < \theta_Y$, or $H_1 : \theta_X \neq \theta_Y$. Note that the algebraic form of the cumulative distribution functions is not assumed known, and so a nonparametric test

1

is used.

To proceed with the test statistic, the differences in observations $d_i = x_i - y_i$ are computed for each of the n pairs, and the signs of these differences are recorded. The *sign of a number* is defined as a plus ("+") if the number is positive and as a minus ("−") if it is negative. If the difference is equal to zero, the sign is undefined. A zero difference, however, is non-informative in this testing procedure. All pairs with zero differences should be eliminated from further consideration, and the total number of pairs valid for the analysis should be reduced accordingly.

Under the null hypothesis that x_i and y_i have the same distribution, their difference d_i is as likely to be positive as negative. Let M denote the total number of positive differences. Then under H_0, M has a binomial distribution with parameter $p = \mathbb{P}(X > Y) = 0.5$. The hypotheses may be written as $H_0 : p = 0.5$ versus $H_1 : p > 0.5$, or $H_1 : p < 0.5$, or $H_1 : p \neq 0.5$. Which alternative hypothesis should be chosen is dictated by the problem at hand. Denote by $Bi(n, 0.5)$ a Binomial random variable with parameters n and 0.5. The P-value is computed as follows:

- For $H_1 : p > 0.5$, P-value $= \mathbb{P}\big(Bi(n, 0.5) \geq M\big) = (0.5)^n \sum_{k=M}^{n} \binom{n}{k}$.

- For $H_1 : p < 0.5$, P-value $= \mathbb{P}\big(Bi(n, 0.5) \leq M\big) = (0.5)^n \sum_{k=0}^{M} \binom{n}{k}$.

- For $H_1 : p \neq 0.5$ and $M \geq n/2$,

$$\text{P-value} = \mathbb{P}\big(Bi(n, 0.5) \geq M \ \text{ or } \ Bi(n, 0.5) \leq n - M\big)$$

$$= (0.5)^n \left[\sum_{k=M}^{n} \binom{n}{k} + \sum_{k=0}^{n-M} \binom{n}{k} \right] = 2(0.5)^n \sum_{k=M}^{n} \binom{n}{k} = (0.5)^{n-1} \sum_{k=M}^{n} \binom{n}{k}.$$

- For $H_1 : p \neq 0.5$ and $M \leq n/2$,

$$\text{P-value} = \mathbb{P}\big(Bi(n, 0.5) \leq M \ \text{ or } \ Bi(n, 0.5) \geq n - M\big)$$

$$= (0.5)^n \left[\sum_{k=0}^{M} \binom{n}{k} + \sum_{k=n-M}^{n} \binom{n}{k} \right] = 2(0.5)^n \sum_{k=0}^{M} \binom{n}{k} = (0.5)^{n-1} \sum_{k=0}^{M} \binom{n}{k}.$$

The next step in the testing procedure is to compare the P-value with the significance level α, which is typically taken as 0.01 or 0.05. In this book, unless mentioned otherwise, α will be assumed equal to 0.05. Finally, if the P-value exceeds α, the null hypothesis is not rejected. Otherwise, it is rejected in favor of the alternative hypothesis. At this point, a conclusion regarding correctness of the tested claim in each particular setting is drawn. It is highly recommended to state the conclusion in plain language addressing non-statisticians.

1.1.2 SAS Implementation

In SAS, the UNIVARIATE procedure may be used to carry out the sign test. The syntax is

```
PROC UNIVARIATE DATA=data_name;
   VAR diff;
RUN;
```

• The variable *diff* contains the difference between matched paired observations. This variable must be computed in a data statement preceding the PROC UNIVARIATE statement.
• The test statistic produced by SAS is $M - n/2$, where the quantity $n/2$ is the expected value of the test statistic M assuming the null hypothesis is true. Incidently, the symbol for the test statistic computed by SAS is also M.
• The P-value in the output is two-sided.

1.1.3 Examples

Example 1.1 An ophthalmologist is testing a new surgical procedure designed to treat *glaucoma*, an elevated intraocular pressure (IOP). He conducts the surgery on one eye of a patient, leaving the other eye as a control. The treated eye is chosen to be the one with the higher pre-surgical value of the IOP. Nine patients are available for the study. The reduction in the IOPs (in millimeters of mercury, mmHg) from pre-surgery to six months following the surgery is recorded. The negative values indicate an actual increase in pressure rather than a decrease. The table below summarizes the results. The last column in the table contains the sign of the difference between the IOP reduction in the treated eye (Tx eye) and that in the control eye (Cx eye).

Patient Number	IOP Reduction in Tx Eye	IOP Reduction in Cx Eye	Sign of Difference
1	0.45	0.38	+
2	1.95	0.90	+
3	1.20	0.70	+
4	0.65	-0.40	+
5	0.98	0.47	+
6	-1.98	-1.30	-
7	1.80	1.34	+
8	-0.76	0.13	-
9	0.56	-0.40	+

The two negative signs in the table indicate that in two cases out of nine the reduction in the IOP was higher in the control eye rather than in the treated eye. To test formally whether the surgery is effective, we have to test $H_1 : p > 0.5$ that states that it is more likely to see a positive sign than a negative one. The test statistic is $M = 7$, the number of positive signs. The P-value is computed as

$$\text{P-value} = \mathbb{P}\big(Bi(9,0.5) \geq 7\big) = (0.5)^9 \sum_{k=7}^{9} \binom{9}{k}$$

$$= (0.5)^9 \left[\binom{9}{7} + \binom{9}{8} + \binom{9}{9} \right] = (0.5)^9 [36 + 9 + 1] = 0.0898.$$

Since P-value > 0.05, the null hypothesis is not rejected at the 5% significance level, and we conclude that the data do not support the efficacy of the surgery.

To implement the sign test in SAS, we submit the code

```
data glaucoma;
   input ID Tx Cx @@;
      diff=Tx-Cx;
   datalines;
1 0.45 0.38 2 1.95 0.90 3 1.20 0.70 4 0.65 -0.50 5 0.98 0.47
6 -1.98 -1.30 7 1.80 1.34 8 -0.76 0.13 9 0.56 -0.40
;

proc univariate data=glaucoma;
   var diff;
run;
```

The relevant SAS output is

```
Test      - Statistic -      - - - -p Value - - - -
Sign      M         2.5      Pr >= |M|   0.1797
```

Note that the test statistic computed by SAS is $M - n/2 = 7 - 9/2 = 7 - 4.5 = 2.5$. Since the default alternative hypothesis in SAS is two-sided, the one-sided P-value for our example can be recovered as $0.1797/2 = 0.0898$. □

Example 1.2 An instructor in an algebra course wishes to test a new approach to teaching algebraic expressions to college freshmen. She sets up a matched paired experiment, where she pairs up two students with similar scores on a midterm exam. Then she randomly selects one student from each pair to be in the intervention group. The other student is assigned to the control group. The intervention group receives the innovative instructions, while the control group receives the standard curriculum instructions. At the end of two weeks, she administers a 100-question quiz and records the number of mistakes. She also records whether the intervention group student's response is higher ("+") or lower ("−") than that of the control group student in the same pair. There are 12 matched pairs in her experiment, a total of 24 freshmen. The data are given in the following table:

Pair Number	Intervention Group	Control Group	Sign of Difference
1	10	26	−
2	22	40	−
3	44	66	−
4	23	55	−
5	8	16	−
6	33	33	0
7	0	8	−
8	8	6	+
9	14	18	−
10	34	14	+
11	2	23	−
12	10	15	−

Since one pair has tied scores, we discard it from the analysis and consider only the 11 pairs with non-tied observations. To test whether the innovative approach is a successful one, we perform a hypotheses testing with the alternative hypothesis $H_1 : p < 0.5$ because a plus sign in this setting indicates a failure. The test statistic is $M = 2$, the number of positive signs, and the corresponding P-value is computed as

$$\text{P-value} = \mathbb{P}\big(Bi(11,0.5) \le 2\big) = (0.5)^{11} \sum_{k=0}^{2} \binom{11}{k}$$

$$= (0.5)^{11} \left[\binom{11}{0} + \binom{11}{1} + \binom{11}{2} \right] = (0.5)^{11} \left[1 + 11 + 55 \right] = 0.0327.$$

Since the P-value is less than 0.05, we reject the null and draw the conclusion that the innovative teaching approach is successful. At the 1% significance level, however, the null hypothesis would not be rejected, disproving the efficacy of the new teaching approach.

We can also verify the analysis by running the following lines of code in SAS:

```
data algebra;
   input pair interv control @@;
      diff=intev-control;
   datalines;
1 10 26 2 22 40 3 44 66 4 23 55 5 8 16 6 33 33
7 0 8 8 8 6 9 14 18 10 34 14 11 2 23 12 10 15
;

proc univariate data=algebra;
   var diff;
run;
```

In the output window, we find the appropriate test statistic and the corresponding two-tailed P-value.

```
Test     -Statistic-      ----p Value----
Sign     M        -3.5    Pr >= |M|   0.0654
```

Finally, we double-check that the SAS test statistic is indeed $M - n/2$ by computing $2 - 11/2 = 2 - 5.5 = -3.5$, and calculate a single-tailed P-value$= 0.0654/2 = 0.0327$, as should be. □

Example 1.3 A doctoral student in real estate management is interested in finding out whether the distribution of home values in a certain residential area has shifted over a two-year period. He acquires monthly house price indices for that area for two consecutive years. The data are (in thousands of dollars):

	House Price Index		Sign of
Month	Year 1	Year 2	Difference
Jan	572	593	−
Feb	572	588	−
Mar	578	586	−
Apr	591	581	+
May	601	576	+
Jun	606	568	+
Jul	602	560	+
Aug	600	555	+
Sep	602	553	+
Oct	604	560	+
Nov	602	566	+
Dec	598	571	+

The student reasonably argues that the index may be subject to seasonal variations, and so he decides to pair the observations for each month. The M-statistic represents the number of positive differences between the indices within each pair, and is equal to 9. Since the direction of the shift in distribution location parameter is of no importance to the student, we choose to test a two-sided alternative hypothesis $H_1 : p \neq 0.5$. In this case, $M \geq n/2$, which results in

$$\text{P-value} = (0.5)^{11} \sum_{k=9}^{12} \binom{12}{k} = (0.5)^{11} \left[\binom{12}{9} + \binom{12}{10} + \binom{12}{11} + \binom{12}{12} \right]$$

$$= (0.5)^{11} [220 + 66 + 12 + 1] = 0.1460.$$

After comparing the P-value to 0.05, we fail to reject the null hypothesis and arrive at the conclusion that there was no evidence for a shift in the distribution location parameter of the house price index over these two years.

We check the result in SAS by running the following code:

```
data home_value;
   input year1 year2 @@;
      diff=year1-year2;
datalines;
572 593 572 588 578 586 591 581 601 576 606 568
602 560 600 555 602 553 604 560 602 566 598 571
;

proc univariate data=home_value;
   var diff;
run;
```

The test statistic and P-value computed by SAS are

```
Test     -Statistic-      ----p Value----
Sign      M          3    Pr >= |M|   0.1460
```

The test statistic $M - n/2 = 9 - 12/2 = 9 - 6 = 3$, and the two-sided P-value coincides with the one calculated by hand. □

1.2 Wilcoxon Signed-Rank Test for Location Parameter for Matched Paired Samples

1.2.1 Testing Procedure

An alternative to the sign test for a matched paired experiment is the Wilcoxon signed-rank test named after a famous American statistician Frank Wilcoxon (1892-1965) who proposed this test in 1945.[1] Note that the sign test studied in the previous section is also typically credited to him.

To carry out the *Wilcoxon signed-rank test*, first we compute the differences $d_i = x_i - y_i$ for each of the n matched pairs. All differences equal to zero should be discarded from the analysis. Then we rank the <u>absolute values</u> of the differences in such a way that the smallest value is assigned a rank of 1, the next smallest, a rank of 2, etc. In case there is a tie between two or more absolute differences, each tied value is assigned the same rank equal to the average of the ranks that would have been assigned if there were no ties, and the next highest absolute difference after the ties are dealt with is assigned the next unused rank. For instance, if two absolute differences are tied for ranks 2 and 3, then each receives a rank of 2.5, and the next highest absolute difference is assigned rank 4.

---

[1]Wilcoxon, F. (1945) Individual comparisons by ranking methods, *Biometrics Bulletin*, **1**, 80-83.

At this stage, the testing branches off in three directions, depending on what the alternative hypothesis is.

- $H_1 : \theta_X < \theta_Y$ is one-sided asserting that the distribution of Y is shifted to the right with respect to the distribution of X. Then the test statistic is the sum of the ranks for the positive differences, T^+, and the null hypothesis should be rejected for small values of T^+, that is, if $T^+ \leq T_0$ where T_0 denotes the critical value described below. Indeed, if, roughly speaking, X is smaller than Y, then the differences $d_i = x_i - y_i$ should be mostly negative and only a few small positive ones are expected.

- $H_1 : \theta_X > \theta_Y$ is one-sided stating that the distribution of Y is shifted to the left with respect to the distribution of X. Then the test statistic is the sum of the ranks for the negative differences, T^-, and the null hypothesis should be rejected for small values of T^-, that is, if $T^- \leq T_0$. Similarly to the above reasoning, if x_i's are larger than y_i's, the differences d_i's are more likely to be positive and larger in magnitude than the negative ones.

On a side note, T^+ and T^- must always add up to the total of all n ranks, that is, $T^- + T^+ = n(n+1)/2$. This formula is useful for verifying correctness of calculations.

- $H_1 : \theta_X \neq \theta_Y$ is two-sided claiming that the distribution of Y is shifted in either direction with respect to the distribution of X. Then the test statistic is $T = \min(T^+, T^-)$, the minimum of T^+ and T^-, and the null is rejected if a small value of T is observed, that is, if $T \leq T_0$.

Note that the test statistic is always chosen in such a way that it is expected to be small under the alternative hypothesis. The critical value T_0 depends on the sample size n, significance level α, and whether the alternative hypothesis is one- or two-tailed. The critical values are tabulated for $\alpha = 0.01$ and 0.05 (see Table A.1 in Appendix A).

1.2.2   Calculation of Critical Values: Example

Here we present an example of how the tabulation is carried out. For simplicity of calculations, we assume that $n = 3$ and there are no tied observations. The ranks of the absolute differences are 1, 2, and 3. The cumulative distribution function of the test statistic T^+ (or T^-, or T) can be computed by listing all possible sequences of signs of the differences. The table below summarizes the results.

Ranks			Statistics		
1	2	3	T^+	T^-	T
+	+	+	6	0	0
−	+	+	5	1	1
+	−	+	4	2	2
+	+	−	3	3	3
−	−	+	3	3	3
−	+	−	2	4	2
+	−	−	1	5	1
−	−	−	0	6	0

As seen from the table, T^+ and T^- have identical distributions. Indeed, it must be so since $T^- = n(n+1)/2 - T^+$. We have that $\mathbb{P}(T^+ \leq 0) = 1/8 = 0.125, \mathbb{P}(T^+ \leq 1) = 2/8 = 0.250, \mathbb{P}(T^+ \leq 2) = 3/8 = 0.375, \mathbb{P}(T^+ \leq 3) = 5/8 = 0.625, \mathbb{P}(T^+ \leq 4) = 6/8 = 0.750, \mathbb{P}(T^+ \leq 5) = 7/8 = 0.875$, and $\mathbb{P}(T^+ \leq 6) = 1$. The cumulative distribution of T is $\mathbb{P}(T \leq 0) = 2/8 = 0.25, \mathbb{P}(T \leq 1) = 4/8 = 0.50, \mathbb{P}(T \leq 2) = 6/8 = 0.75$, and $\mathbb{P}(T \leq 3) = 1$.

Note that none of these tail probabilities are anywhere near the typically chosen critical values of 0.01 or 0.05. Only starting at $n = 5$, such small probabilities can be seen. Hence Table A.1 starts with $n = 5$.

1.2.3 SAS Implementation

To conduct the signed-rank test in SAS, use the PROC UNIVARIATE statement exactly as we did for the sign test. We repeat it here for convenience.

```
PROC UNIVARIATE DATA=data_name;
VAR diff;
RUN;
```

- SAS outputs $S = T^+ - n(n+1)/4 = n(n+1)/4 - T^-$ as the test statistic. Here $n(n+1)/4$ is the expected value of T^+ under the null hypothesis. The P-value is given for a two-sided alternative hypothesis.
- For $n > 20$, the P-value is computed based on asymptotic distribution of the test statistic S, according to which $S\sqrt{(n-1)/(n\mathbb{V}ar(S) - S^2)}$ has a t-distribution with $n - 1$ degrees of freedom. Here the variance of S is $\mathbb{V}ar(S) = n(n+1)(2n+1)/24$ (more generally, if ties are present, the variance is $n(n+1)(2n+1)/24 - \sum_{j=1}^{m}(T_j^3 - T_j)/48$ where $T_j, j = 1, \ldots, m$, is the size of the jth group of tied observations).

1.2.4   Examples

Example 1.4 To perform the signed-rank test for the data in Example 1.1, we assign the ranks to the absolute differences. The results are summarized in the following table:

Patient Number	IOP Reduction in Tx Eye	IOP Reduction in Cx Eye	Difference	Rank
1	0.45	0.38	0.07	1
2	1.95	0.90	1.05	8
3	1.20	0.70	0.50	3
4	0.65	−0.50	1.15	9
5	0.98	0.47	0.51	4
6	−1.98	−1.30	−0.68	5
7	1.80	1.34	0.46	2
8	−0.76	0.13	−0.89	6
9	0.56	−0.40	0.96	7

For the hypothesis under investigation $H_1 : \theta_{Tx} > \theta_{Cx}$, the rejection region consists of sufficiently small sums of the ranks for the negative differences. The test statistic is $T^- = 5 + 6 = 11$. From Table A.1, for $n = 9$, the critical value for a one-sided test at the 5% significance level is $T_0 = 8$. Since $T^- > T_0$, the null hypothesis is not rejected, leading to the conclusion that the surgery is not effective. The same conclusion was reached in Example 1.1.

Next, we run the same lines of SAS code as in Example 1.1. The output pertained to the signed-rank test is

```
Test              - Statistic -       - - - - p Value - - - -
Signed Rank      S        11.5      Pr >= |S|    0.2031
```

Note that the test statistic computed by SAS is $S = n(n+1)/4 - T^- = 9(9+1)/4 - 11 = 22.5 - 11 = 11.5$. The one-sided P-value for this example is $0.2031/2 = 0.1015$. Since it is larger than 0.05, the null hypothesis should not be rejected. □

Example 1.5 For the data in Example 1.2, the differences and their ranks are calculated as follows:

Pair Number	Intervention Group	Control Group	Difference	Rank
1	10	26	−16	6
2	22	40	−18	7
3	44	66	−22	10
4	23	55	−32	11
5	8	16	−8	4.5
6	33	33	0	N/A
7	0	8	−8	4.5
8	8	6	2	1
9	14	18	−4	2
10	34	14	20	8
11	2	23	−21	9
12	10	15	−5	3

The case with the identical observations is removed from further consideration, leaving the valid sample size $n = 11$. There are two tied differences which are assigned the rank of 4.5 each, a half way between the ranks 4 and 5 that would have been assigned to them if they were not tied. The alternative hypothesis for this setting is $H_1 : \theta_{interv} < \theta_{control}$. The test statistic is $T^+ = 1 + 8 = 9$ since positive differences are not welcome if the null is to be rejected. As seen in Table A.1, the critical value for the one-tailed test at $\alpha = 0.05$ is equal to 13, thus indicating the rejection of the null hypothesis at the 5% level of significance. Note that the critical value corresponding to the 1% significance level is 7, so we fail to reject the null hypothesis at $\alpha = 0.01$. The conclusion is that the intervention is effective when judging at the 5% level of significance. It is not, however, effective at the 1% significance level. The same result was obtained previously in Example 1.2.

Running the SAS code given in Example 1.2 produces the following output:

```
Test            -Statistic-      ----p Value----
Signed Rank      S       -24     Pr >= |S|    0.0313
```

The test statistic computed by SAS is $S = T^+ - n(n+1)/4 = 9 - 11(11+1)/4 = 9 - 33 = -24$. The one-sided P-value is derived as 0.0313/2=0.0157. Note that this P-value is indeed smaller than 0.05 but larger than 0.01. □

Example 1.6 Referring back to the data set in Example 1.3, we assign the ranks to the absolute differences as follows:

	Home Value Index			
Month	Year 1	Year 2	Difference	Rank
Jan	572	593	−21	4
Feb	572	588	−16	3
Mar	578	586	−8	1
Apr	591	581	10	2
May	601	576	25	5
Jun	606	568	38	8
Jul	602	560	42	9
Aug	600	555	45	11
Sep	602	553	49	12
Oct	604	560	44	10
Nov	602	566	36	7
Dec	598	571	27	6

For the two-sided alternative hypothesis $H_1 : \theta_{Year1} \neq \theta_{Year2}$, the test statistic is $T = \min(T^+, T^-) = T^- = 4 + 3 + 1 = 8$. As found in Table A.1, the critical value that corresponds to $n = 12$, two-tailed test, and $\alpha = 0.05$ is 13 (for $\alpha = 0.01$, $T_0 = 7$), hence, the null should be rejected at the 5% but not 1% significance level. Thus, at the level of 0.05, there is enough evidence in the data to conclude that there is a shift in the distribution of home values. This conclusion cannot be drawn at the level of 0.01. Note that for these data, the sign test and signed-rank test lead to different conclusions.

The code from Example 1.3 gives the following result:

```
Test            - Statistic -      ----p Value----
Signed Rank     S          31      Pr >= |S|   0.0122
```

The output test statistic is indeed $S = n(n+1)/4 - T^- = 12(12+1)/4 - 8 = 39 - 8 = 31$. The P-value of 0.0122 is for a two-tailed alternative hypothesis, supporting our claim that H_0 is rejected at the 0.05 but not at the 0.01 significance level. □

1.3 Wilcoxon Rank-Sum Test for Location Parameter for Two Independent Samples

If measurements for two independent samples are observed and the data may be assumed normal, then a two-sample t-test is carried out. If normality doesn't hold, the Wilcoxon rank-sum test may be used as a nonparametric alternative. This test and the signed-rank test (see Section 1.2) were proposed by Frank Wilcoxon in the same paper.[1]

---

[1] Wilcoxon, F. (1945) Individual comparisons by ranking methods, *Biometrics Bulletin*, **1**, 80-83.

1.3.1 Test Procedure

Suppose observations $x_1, \ldots, x_{n_1}$ and $y_1, \ldots, y_{n_2}$ are available. It is assumed that the two samples are drawn from independent populations. To test a hypothesis regarding relative position of the location parameters of the underlying distributions, the *Wilcoxon rank-sum test* may be conducted. The following are the steps of the testing procedure. First, the samples are pooled and the observations are ranked so that the smallest observation receives the rank of 1. If two or more observations are tied, all of them are assigned the same rank which equals to the average of the ranks that the observations should have gotten if they were not tied. Then the test statistic W is computed as the sum of the ranks assigned to the observations in the smaller sample. For definiteness, we will assume that the first sample, the sample of size n_1, is smaller. This can always be achieved by renaming samples. If the samples have equal sizes, we take the first sample.

The next step depends on the type of the alternative hypothesis under investigation.

- $H_1 : \theta_X < \theta_Y$ is one-sided asserting that the distribution of Y is shifted to the right with respect to the distribution of X. Under this alternative hypothesis, the test statistic is expected to be small, thus the null hypothesis should be rejected for small values of W, that is, if $W \leq W_L$ where W_L denotes the lower critical value introduced below.

- $H_1 : \theta_X > \theta_Y$ is one-sided stating that the distribution of Y is shifted to the left with respect to the distribution of X. If this alternative is true, then the test statistic tends to be large. Hence, the null hypothesis should be rejected for large values of W, that is, if $W \geq W_U$ where W_U is the upper critical value defined below.

- $H_1 : \theta_X \neq \theta_Y$ is two-sided with the direction of the location shift not specified. Under this alternative, the test statistic should be either small or large, therefore, the null hypothesis is rejected if W falls outside of the critical interval (W_L, W_U).

The lower and upper critical values W_L and W_U depend on the sample sizes n_1 and n_2, significance level α (typically chosen as 0.01 or 0.05), and whether the alternative is one- or two-tailed. The critical values are tabulated (see Table A.2 in Appendix A).

1.3.2 Calculation of Critical Values: Example

To see how the critical values are computed, consider an example with $n_1 = n_2 = 2$, and observations $x_1, x_2, y_1,$ and y_2. There is a total of $4! = 24$ ways to assign ranks 1 through 4 to these observations (we rule out ties). The table below shows all possibilities of rank assignment, the test statistic W (the sum of the ranks given to x_1 and x_2), and the corresponding probability.

Ranks				Test Statistic	Probability
x_1	x_2	x_3	x_4		
1	2	3	4		
1	2	4	3	$1+2=3$	$4/24=0.1667$
2	1	3	4		
2	1	4	3		
1	3	2	4		
1	3	4	2	$1+3=4$	$4/24=0.1667$
3	1	2	4		
3	1	4	2		
1	4	2	3		
1	4	3	2		
4	1	2	3		
4	1	3	2	$1+4$	$8/24=0.3333$
2	3	1	4	$=2+3=5$	
2	3	4	1		
3	2	1	4		
3	2	4	1		
2	4	1	3		
2	4	3	1	$2+4=6$	$4/24=0.1667$
4	2	1	3		
4	2	3	1		
3	4	1	2		
3	4	2	1	$3+4=7$	$4/24=0.1667$
4	3	1	2		
4	3	2	1		

The lower-tail probabilities are $\mathbb{P}(W \leq 3) = 0.1667$, $\mathbb{P}(W \leq 4) = 0.3333$, $\mathbb{P}(W \leq 5) = 0.6667$, $\mathbb{P}(W \leq 6) = 0.8333$, and $\mathbb{P}(W \leq 7) = 1$. For larger sample sizes, some of these probabilities become less than or equal to 0.05 (or 0.01). The corresponding percentile is taken as the lower critical value. The upper critical values are computed analogously.

1.3.3   SAS Implementation

SAS outputs a test statistic and P-value for the rank-sum test. To request these, procedure NPAR1WAY is used. As the name suggests, this procedure performs a variety of nonparametric tests for one-way classification (that is, for one variable). The syntax is

```
PROC NPAR1WAY DATA=data_name WILCOXON;
     CLASS sample_name;
        VAR variable_name;
     EXACT;
```

RUN;

- The CLASS statement identifies samples that are being compared.
- The WILCOXON option requests the Wilcoxon rank-sum test to be performed. In SAS it is termed the *Wilcoxon two-sample test*.
- The EXACT option requests an *exact test* for the specified statistic, that is, an exact P-value is computed via listing all possibilities.
- The P-value based on asymptotic normality of the test statistic is also computed. For large $n_1 \leq n_2$, the test statistic is approximately normally distributed with mean $n_1(n+1)/2$ and variance $n_1 n_2(n+1)/12$ where $n = n_1 + n_2$ (if corrected for ties, the

variance is $n_1 n_2(n+1)/12 - n_1 n_2/[12n(n-1)] \sum_{j=1}^{m} T_j(T_j^2 - 1)$ with $T_1, \ldots, T_m$ denot-

ing the respective sizes of m groups of tied observations).

1.3.4   Examples

Example 1.7 To investigate the effectiveness of a student learning assistance program in a college, ten names were drawn randomly from the list of seniors who had completed the program, and, independently, ten names were randomly selected from among those seniors who did not participate in the program. The GPAs for all the students were recorded. The data are

Program Yes	Rank	Program No	Rank
3.98	18	3.42	7
3.45	9.5	2.56	2
3.66	13	2.00	1
3.78	14.5	3.19	6
3.90	16	3.00	4
4.00	19.5	3.56	11.5
3.78	14.5	3.56	11.5
3.12	5	4.00	19.5
3.45	9.5	2.78	3
3.97	17	3.44	8

The rank-sum test statistic $W = 136.5$ is the sum of the ranks for program completers. If the program is effective, then W should be a large number. From Table A.2, the upper critical value for a one-tailed alternative $H_1 : \theta_{yes} > \theta_{no}$ at $n_1 = n_2 = 10$ and $\alpha = 0.05$ is $W_U = 128$, indicating that the null hypothesis is to be rejected at the 0.05 significance level. Note that the upper critical value for $\alpha = 0.01$ is 136 implying that the null hypothesis should be rejected even at the 1% significance level. The conclusion is that the program is in fact effective.

In SAS we run the following code:

```
data learning_program;
      input program $ GPA @@;
   datalines;
yes 3.98 no 3.42 yes 3.45 no 2.56 yes 3.66
no 2.00 yes 3.78 no 3.19 yes 3.90 no 3.00
yes 4.00 no 3.56 yes 3.78 no 3.56 yes 3.12
no 4.00 yes 3.45 no 2.78 yes 3.97 no 3.44
;

proc npar1way data=learning_program wilcoxon;
   class program;
      var GPA;
   exact;
run;
```

The output is

```
   Wilcoxon Two-Sample Test
 Statistic (S)        136.5000
 Exact Test
 One-Sided Pr >= S     0.0078
```

The output test statistic is the same as the one computed by hand. The P-value is below 0.01, which validates our conclusion. □

Example 1.8 The performance of a new wound healing bioimplant is studied retro-spectively. In a medical center, soft tissue in foot and ankle was repaired per standard procedure and augmented with the bioimplant in six subjects (Tx group), whereas it was repaired but not augmented in nine subjects (Cx group). Time to full weight bearing (in weeks) is available for all subjects. The research hypothesis is that the bioimplant is effective. The observations and their ranks are given in the following table:

Tx group	3	4	6	6	8	8			
Rank	1	3.5	8	8	12	12			
Cx group	4	4	4	5	6	7	8	10	12
Rank	3.5	3.5	3.5	6	8	10	12	14	15

The sum of the ranks for the smaller sample, the treatment group, is $W = 44.5$. The sample sizes are $n_1 = 6$ and $n_2 = 9$. The alternative hypothesis $H_1 : \theta_{Tx} < \theta_{Cx}$ is lower-tailed and hence small values of the test statistic are favorable. The lower 0.05-level critical value as found in Table A.2 is $W_L = 33$, implying that the null hypothesis cannot be rejected, and, therefore, there is no evidence in the data that the bioimplant is effective.

The SAS output supports this conclusion. The code is

```
data bioimplant;
  input group $ timeFWB @@; /*timeFWB=time to full weight bearing*/
    datalines;
Tx 3 Tx 4 Tx 6 Tx 6 Tx 8 Tx 8
Cx 4 Cx 4 Cx 4 Cx 5 Cx 6 Cx 7 Cx 8 Cx 10 Cx 12
;

proc npar1way data=bioimplant wilcoxon;
    class group;
        var timeFWB;
    exact;
run;
```

The output states

```
   Wilcoxon Two-Sample Test
   Statistic (S)        44.5000
   Exact Test
   One-Sided Pr <= S    0.3491
```

The P-value exceeds 0.05, thus we fail to reject the null hypothesis.   □

Example 1.9 A survey was conducted on men and women aged 70-79 years who live in a small elderly community. One of the variables recorded in the survey was self-efficacy. This variable was measured on the five-point Likert scale (1=not self-efficient, 2=somewhat self-efficient, 3=mostly self-efficient, 4=self-efficient, 5=highly self-efficient). Valid responses were obtained for six men and seven women. The observations and their corresponding ranks are shown in the table below. The objective is to test whether there is a difference in self-efficacy by gender.

Men	Rank	Women	Rank
4	6	5	11.5
5	11.5	5	11.5
4	6	4	6
3	1.5	4	6
4	6	4	6
3	1.5	5	11.5
		4	6

The rank-sum test statistic is $W = 32.5$, the sum of the ranks for men since they comprise a smaller sample. From Table A.2, the critical values for $n_1 = 6, n_2 = 7, \alpha = 0.05$, and a two-sided alternative $H_1 : \theta_{men} \neq \theta_{women}$ are $W_L = 27$ and $W_U = 57$. Since the test statistic falls between the critical limits, the null hypothesis cannot be rejected, yielding the conclusion that there is no difference in self-efficacy between men and women.

This result can be verified in SAS by running the following lines of code:

```
data elderly_study;
      input gender $ self_efficacy @@;
   datalines;
M 4 M 5 M 4 M 3 M 4 M 3
F 5 F 5 F 4 F 4 F 4 F 5 F 4
;

proc npar1way data=elderly_study wilcoxon;
   class gender;
      var self_efficacy;
   exact;
run;
```

The output is

```
      Wilcoxon Two-Sample Test
 Statistic (S)              32.5000
 Exact Test
 Two-Sided Pr >= |S-Mean|   0.2284
```

The test statistic is the same as computed by hand, and the two-sided P-value is 0.2284, indicating the failure to reject the null hypothesis. □

1.4 Ansari-Bradley Test for Scale Parameter for Two Independent Samples

If observations from two independent samples are normally distributed, and a test for equality of variances is to be conducted, then the standard F-test handles this situation well. In case the normality assumption is violated, and it is desired to test equality of *scale parameters* determining the spread of the probability distributions, then the nonparametric *Ansari-Bradley test* may be implemented. Its use is validated, however, only if the location parameters of the probability distributions are the same.

This test was first published in 1960 by Ralph Allen Bradley (1923-2001), a Canadian-American statistician, and a then Ph.D. student Abdur Rahman Ansari.[1]

1.4.1 Test Procedure

Suppose n_1 observations of variable X and n_2 observations of variable Y are available. These samples come from two independent populations. It is assumed that the two cumulative distribution functions, F_X and F_Y, have equal location parameters, say, θ,

[1] Ansari, A.R. and Bradley, R.A. (1960) Rank sum tests for dispersion. *The Annals of Mathematical Statistics*, **31**, 1174-1189.

but different scale parameters, say, η_1 and η_2, respectively. Let $\gamma = \eta_2/\eta_1$ denote the ratio of the two scale parameters. The hypotheses of interest may be easily expressed in terms of γ. As always, the null $H_0 : \gamma = 1$ asserts that the scale parameters are equal. The alternative hypothesis may be taken as either one-sided lower-tailed $H_1 : \gamma < 1$, or one-sided upper-tailed $H_1 : \gamma > 1$, or two-sided $H_1 : \gamma \neq 1$, depending on particular setting.

To compute the test statistic, first we arrange the combined $n = n_1 + n_2$ observations in increasing order, and then assign ranks in such a way that the smallest and the largest are each given a rank of 1, the second smallest and the second largest both get a rank of 2, etc. If n is an even number, the complete set of ranks will be $1, 2, \ldots, n/2 - 1, n/2, n/2, n/2 - 1, \ldots, 2, 1$. If n is odd, the complete set of ranks will be $1, 2, \ldots, (n-1)/2, (n+1)/2, (n-1)/2, \ldots, 2, 1$. If tied observations are present, we assign to them the same rank which equals to the average of the ranks that would have been normally assigned to these observations were they not tied. Finally, we compute the test statistic C as the sum of ranks assigned to the smaller sample. Without loss of generality, we can assume that the sample of size n_1 is smaller. We can always state the alternative hypothesis in accordance with this assumption. If the samples have equal sizes, we sum the ranks of the observations in the first sample. Further, the three cases are distinguished:

• Under $H_1 : \gamma < 1$, asserting that observations from the second population are not so spread out as the ones in the first population, the test statistic is expected to be small. Indeed, if X-observations are spread out more, they are more likely to appear on the flanks of the ordered set, and are given smaller ranks. Thus, the sum of these ranks is expected to be small. The null hypothesis is rejected in favor of the alternative, if the test statistic C is smaller than or equal to the lower critical value C_L (see below) for a one-tailed test at a specified α-level.

• Under $H_1 : \gamma > 1$, claiming that observations from the second population are more spread out than the ones in the first population, the test statistic is expected to be large. Therefore, the null hypothesis is rejected, if the test statistic C exceeds or is equal to the upper critical value C_U (see below) for a one-tailed test at a desired level of significance.

• Under the two-sided $H_1 : \gamma \neq 1$, observations from the two populations are not equally spread out. For the null hypothesis to be rejected, the test statistic C should lie outside of the open interval between the lower critical value C_L and the upper critical value C_U for a two-tailed test for a given α.

The critical values C_L and C_U have been pre-calculated for various sample sizes $n_1 \leq n_2$, different α-levels, and for one- and two-sided tests. The critical values for $\alpha = 0.01$ and 0.05 are summarized in Table A.3 in Appendix A.

1.4.2   Calculation of Critical Values: Examples

• Consider the case $n_1 = n_2 = 2$, so that $n = n_1 + n_2 = 4$ is an even number. Assume no ties in the data. The sequence of ranks is 1, 2, 2, 1. There are six possibilities of rank assignments. They are given in the following table, along with the corresponding values of the test statistic C (the sum of the ranks for x_1 and x_2), and their probabilities:

Ranks 1 2 2 1	Test Statistic	Probability
$x_1\, y_1\, y_2\, x_2$	$1 + 1 = 2$	$1/6 = 0.1667$
$x_1\, x_2\, y_1\, y_2$	$1 + 2 = 3$	
$x_1\, y_1\, x_2\, y_2$	$1 + 2 = 3$	
$y_1\, x_1\, y_2\, x_2$	$2 + 1 = 3$	$4/6 = 0.6667$
$y_1\, y_2\, x_1\, x_2$	$2 + 1 = 3$	
$y_1\, x_1\, x_2\, y_2$	$2 + 2 = 4$	$1/6 = 0.1667$

The cumulative probabilities for the test statistic are $\mathbb{P}(C \leq 2) = 1/6 = 0.1667$, $\mathbb{P}(C \leq 3) = 1/6 + 4/6 = 5/6 = 0.8333$, and $\mathbb{P}(C \leq 4) = 1$.

• Consider now the case $n_1 = 2$, and $n_2 = 3$. Assume the absence of tied observations. The complete set of ranks to be assigned is 1, 2, 3, 2, 1. There are 10 possibilities of the rank assignment, which are presented in the following table:

Ranks 1 2 3 2 1	Test Statistic	Probability
$x_1\, y_1\, y_2\, y_3\, x_2$	$1 + 1 = 2$	$1/10 = 0.1$
$x_1\, x_2\, y_1\, y_2\, y_3$	$1 + 2 = 3$	
$x_1\, y_1\, y_2\, x_2\, y_3$	$1 + 2 = 3$	
$y_1\, x_1\, y_2\, y_3\, x_2$	$2 + 1 = 3$	$4/10 = 0.4$
$y_1\, y_2\, y_3\, x_1\, x_2$	$2 + 1 = 3$	
$x_1\, y_1\, x_2\, y_2\, y_3$	$1 + 3 = 4$	
$y_1\, x_1\, y_2\, x_2\, y_3$	$2 + 2 = 4$	$3/10 = 0.3$
$y_1\, y_2\, x_1\, y_3\, x_2$	$3 + 1 = 4$	
$y_1\, x_1\, x_2\, y_2\, y_3$	$2 + 3 = 5$	
$y_1\, y_2\, x_1\, x_2\, y_3$	$3 + 2 = 5$	$2/10 = 0.2$

The lower-tail probabilities for the test statistic are $\mathbb{P}(C \leq 2) = 0.1$, $\mathbb{P}(C \leq 3) = 0.5$, $\mathbb{P}(C \leq 4) = 0.8$, and $\mathbb{P}(C \leq 5) = 1$.

For larger values of $n_1 \leq n_2$, the lower- and upper-tail probabilities become smaller than 0.05 (or 0.01), yielding lower and upper critical values for tabulation.

1.4.3   SAS Implementation

To run the Ansari-Bradley test in SAS, the NPAR1WAY procedure can be used:

```
PROC NPAR1WAY DATA=data_name AB;
   CLASS sample_name;
      VAR variable_name;
   EXACT;
RUN;
```

- Here the option AB stands for the Ansari-Bradley test.
- Besides the exact P-value, SAS also outputs an approximate P-value based on asymptotic normality of the test statistic C, the mean and variance of which are computed as follows. Assuming $n_1 \le n_2$, if $n = n_1 + n_2$ is even, then $\mathbb{E}(C) = n_1(n+2)/4$ and $\mathbb{V}ar(C) = n_1 n_2 (n^2 - 4)/[48(n-1)]$ (more generally, $\mathbb{V}ar(C) = n_1 n_2 \left[16 \sum_{j=1}^{m} T_j r_j^2 - n(n+2)^2\right]/[16n(n-1)]$ in case of m groups of tied observations of sizes $T_1,\ldots,T_m$ with associated ranks $r_1,\ldots,r_m$, respectively); and if $n = n_1 + n_2$ is odd, then $\mathbb{E}(C) = n_1(n+1)^2/(4n)$ and $\mathbb{V}ar(C) = n_1 n_2 (n+1)(n^2+3)/(48n^2)$ (in case of ties, $\mathbb{V}ar(C) = n_1 n_2 \left[16n \sum_{j=1}^{m} T_j r_j^2 - (n+1)^4\right]/[16n^2(n-1)]$).

1.4.4　Examples

Example 1.10 A researcher is interested in comparing two standard psychological tests. The first test has been around for decades, while the second one was introduced relatively recently. The researcher hypothesizes that the newer test is more precise than the older one. He collects scores on the first test for five subjects, and on the second one, for seven subjects. The scales of the tests differ, so the data have to be re-scaled to reflect percentages of the maximum possible score. The re-scaled data are

Older Test	72	64	34	78	87		
Ranks	5.5	3	1	5.5	2		
Newer Test	80	72	94	68	57	78	82
Ranks	4	5.5	1	4	2	5.5	3

The test statistic is the sum of the ranks given to the older test scores (the smaller sample). It is $C = 17$. The alternative hypothesis in this situation is $H_1 : \gamma < 1$ since we want to find out whether in the second population the distribution scale parameter is smaller (this is the meaning of a more precise psychological test). From Table A.3, for $n_1 = 5$, $n_2 = 7$, and $\alpha = 0.05$, the lower critical value $C_L = 12$. The test statistic exceeds this critical value, thus we fail to reject the null hypothesis, and say that there is not sufficient evidence to suggest that the newer test is more precise.

To back up this conclusion with SAS output, we run the following code:

```
data psy_tests;
     input test $ score @@;
   datalines;
```

```
older 72 older 64 older 34 older 78 older 87
newer 80 newer 72 newer 94 newer 68 newer 57 newer 78 newer 82
;

proc npar1way data=psy_tests ab;
    class test;
        var score;
    exact;
run;
```

The output is

```
Ansari-Bradley Two-Sample Test
Statistic (S)              17.0000
Exact Test
One-Sided Pr <= S          0.4747
```

The large P-value supports our conclusion. ☐

Example 1.11 In Example 1.8, we concluded that the location parameters of the distributions of the time to full weight bearing are equal for the treated and control patients. The next advancement may be to find out whether the dispersion of the time to full weight bearing for the treated patients is smaller than that for the control. To answer this question, first assign the ranks according to the Ansari-Bradley procedure. The ranks are

Tx group	3	4	6	6	8	8			
Rank	1	3.5	7.33	7.33	4	4			
Cx group	4	4	4	5	6	7	8	10	12
Rank	3.5	3.5	3.5	6	7.33	6	4	2	1

The test statistic $C = 27.16$ is the sum of the ranks that correspond to the observations in the smaller sample (Tx group). The tested alternative hypothesis is upper-tailed, $H_1 : \gamma > 1$. Under this alternative, the test statistic should be large, therefore, we compare C to the upper critical value $C_U = 34$ for $n_1 = 6$, $n_2 = 9$, and $\alpha = 0.05$, as found in Table A.3. Since the observed test statistic is smaller than the upper critical value, there is not enough evidence to reject the null, and the conclusion is that the dispersion of the time to full weight bearing for the treatment group is not smaller than that for the control group.

The code in SAS is

```
proc npar1way data=bioimplant ab;
    class group;
        var timeFWB;
    exact;
```

```
run;
```

The relevant output is

```
Ansari-Bradley Two-Sample Test
Statistic (S)           27.1667
Exact Test
One-Sided Pr >= S         0.3652
```

The P-value is larger than 0.05, which is in sync with the above conclusion. □

Example 1.12 Consider the data in Example 1.9. The conclusion drawn from those data were that there is no difference between location parameters of the distribution of self-efficacy scores for men and women. Next, we wish to know whether the variabilities in the scores differ by gender. We apply the Ansari-Bradley test with the following ranks:

Men	Rank	Women	Rank
4	5.14	5	2.5
5	2.5	5	2.5
4	5.14	4	5.14
3	1.5	4	5.14
4	5.14	4	5.14
3	1.5	5	2.5
		4	5.14

The test statistic is the sum of the ranks in the smaller sample, $C = 20.92$. As seen in Table A.3, the 5% critical values for the two-tailed alternative $H_1 : \gamma \neq 1$ that correspond to the sample sizes $n_1 = 6$ and $n_2 = 7$ are $C_L = 16$ and $C_U = 30$. Since the test statistic is between the critical values, we fail to reject the null hypothesis and state that the variabilities do not differ by gender.

Running the following code in SAS supports the conclusion.

```
proc npar1way data=elderly_study ab;
   class gender;
      var self_efficacy;
   exact;
run;
```

The output is

```
Ansari-Bradley Two-Sample Test
Statistic (S)           20.9286
Exact Test
Two-Sided Pr >= |S-Mean|    0.6533
```

The test statistic equals the one computed by hand up to a round-off error. The P-value is much larger than 0.05, leading to the same conclusion as above.   □

1.5   Kolmogorov-Smirnov Test for Equality of Distributions

Suppose measurements are taken for two samples drawn from two independent populations. The question of interest is whether the underlying distributions for these measurements in the respective populations are equal. A nonparametric *Kolmogorov-Smirnov test* may be conducted in this case. This test was developed by Soviet mathematicians Andrey N. Kolmogorov (1903-1987) and Nikolai V. Smirnov (1900-1966).[1,2]

1.5.1   Testing Procedure

Let $x_1, \ldots, x_{n_1}$ and $y_1, \ldots, y_{n_2}$ be two independent random samples from populations with continuous cumulative distribution functions F_X and F_Y, respectively. We would like to assess whether these are the same functions, that is, we would like to test the null hypothesis

$$H_0: \; F_X(t) = F_Y(t) \; \text{ for all } \; t$$

against one of the three alternative hypotheses:

$$H_1: \; F_X(t) > F_Y(t) \; \text{ for some } \; t,$$

or

$$H_1: \; F_X(t) < F_Y(t) \; \text{ for some } \; t,$$

or

$$H_1: \; F_X(t) \neq F_Y(t) \; \text{ for some } \; t.$$

As always, a particular context dictates whether the alternative should be upper-tailed, lower-tailed or two-tailed. To apply the Kolmogorov-Smirnov test, first compute the respective *empirical distribution functions* $\hat{F}_X(t)$ and $\hat{F}_Y(t)$ for the two samples. These are defined for any t as

$$\hat{F}_X(t) = \frac{\# \text{ of } x's \leq t}{n_1} = \frac{1}{n_1} \sum_{i=1}^{n_1} \mathbb{I}\{x_i \leq t\},$$

and

$$\hat{F}_Y(t) = \frac{\# \text{ of } y's \leq t}{n_2} = \frac{1}{n_2} \sum_{i=1}^{n_2} \mathbb{I}\{y_i \leq t\}.$$

---

[1]Kolmogorov, A. N. (1933) On the empirical determination of a distribution function (in Italian), *Giornale dell'Instituto Italiano degli Attuari*, **4**, 83-91.

[2]Smirnov, N. V. (1939) On the estimation of the discrepancy between empirical curves of distribution for two independent samples (in Russian), *Bulletin of Moscow University*, **2**, 3-16.

In the above, the function $\mathbb{I}\{A\}$ denotes the *indicator function* of a statement A, which is equal to one if the statement is true and zero, otherwise.

Next, the test statistic is calculated. It represents the greatest *discrepancy* (or the greatest *vertical distance*) between the two empirical distribution functions.

- For the upper-tailed alternative $H_1 : F_X(t) > F_Y(t)$ for some t, the test statistic is $D^+ = \max_t \left(\hat{F}_X(t) - \hat{F}_Y(t)\right)$.

- For the lower-tailed alternative $H_1 : F_X(t) < F_Y(t)$ for some t, the test statistic is $D^- = \max_t \left(\hat{F}_Y(t) - \hat{F}_X(t)\right)$.

- For the two-tailed alternative $H_1 : F_X(t) \neq F_Y(t)$ for some t, the test statistic is $D = \max_t \left|\hat{F}_X(t) - \hat{F}_Y(t)\right|$.

In practice, the test statistic is computed using the fact that the empirical distributions are step functions changing values only at the sample points. Thus, if we denote the pooled ordered data points by $z_{(1)}, \ldots, z_{(n_1+n_2)}$, we can write the computational formulas for the test statistics as:

$$D^+ = \max_{i=1,\ldots,n_1+n_2} \left(\hat{F}_X(z_{(i)}) - \hat{F}_Y(z_{(i)})\right),$$

$$D^- = \max_{i=1,\ldots,n_1+n_2} \left(\hat{F}_Y(z_{(i)}) - \hat{F}_X(z_{(i)})\right),$$

and

$$D = \max_{i=1,\ldots,n_1+n_2} \left|\hat{F}_X(z_{(i)}) - \hat{F}_Y(z_{(i)})\right|.$$

Finally, the test statistic is compared to the appropriate critical value. Critical values are tabulated (see Table A.4 in Appendix A) for different combinations of the sample sizes $n_1 \leq n_2$ and significance levels α. The decision rule is to reject the null hypothesis if the test statistic is strictly larger than the critical value, and fail to reject the null, otherwise.

An important remark is in order. The critical values in Table A.4 were calculated under the assumption of no ties in the data. If tied observations are present, the Kolmogorov-Smirnov test procedure remains valid, and the same critical values may be used. The resulting test would have the significance level not exceeding the nominal level α.

1.5.2   Calculation of Critical Values: Example

Consider an example of two observations in the first sample and three observations in the second sample. No ties are assumed to be present in the data. Without loss of generality, the five observations can be taken as 1, 2, 3, 4, and 5. The total number of distinct permutations is $\binom{5}{2} = 10$.

The Kolmogorov-Smirnov two-sided test statistic $D = \max |\hat{F}_1 - \hat{F}_2|$ is computed for each permutation. The procedure is summarized below.

Sample 1	Sample 2	Obs	$\hat{F}_1$	$\hat{F}_2$	$\|\hat{F}_1 - \hat{F}_2\|$	D
		1	1/2	0	1/2	
		2	1	0	1	
1 2	3 4 5	3	1	1/3	2/3	1
		4	1	2/3	1/3	
		5	1	1	0	
		1	1/2	0	1/2	
		2	1/2	1/3	1/6	
1 3	2 4 5	3	1	1/3	2/3	2/3
		4	1	2/3	1/3	
		5	1	1	0	
		1	1/2	0	1/2	
		2	1/2	1/3	1/6	
1 4	2 3 5	3	1/2	2/3	1/6	1/2
		4	1	2/3	1/3	
		5	1	1	0	
		1	1/2	0	1/2	
		2	1/2	1/3	1/6	
1 5	2 3 4	3	1/2	2/3	1/6	1/2
		4	1/2	1	1/2	
		5	1	1	0	
		1	0	1/3	1/3	
		2	1/2	1/3	1/6	
2 3	1 4 5	3	1	1/3	2/3	2/3
		4	1	2/3	1/3	
		5	1	1	0	
		1	0	1/3	1/3	
		2	1/2	1/3	1/6	
2 4	1 3 5	3	1/2	2/3	1/6	1/3
		4	1	2/3	1/3	
		5	1	1	0	
		1	0	1/3	1/3	
		2	1/2	1/3	1/6	
2 5	1 3 4	3	1/2	2/3	1/6	1/2
		4	1/2	1	1/2	
		5	1	1	0	

(To be continued...)

(Continued...)

Sample 1	Sample 2	Obs	$\hat{F}_1$	$\hat{F}_2$	$|\hat{F}_1 - \hat{F}_2|$	D
		1	0	1/3	1/3	
		2	0	2/3	2/3	
3 4	1 2 5	3	1/2	2/3	1/6	2/3
		4	1	2/3	1/3	
		5	1	1	0	
		1	0	1/3	1/3	
		2	0	2/3	2/3	
3 5	1 2 4	3	1/2	2/3	1/6	2/3
		4	1/2	1	1/2	
		5	1	1	0	
		1	0	1/3	1/3	
		2	0	2/3	2/3	
4 5	1 2 3	3	0	1	1	1
		4	1/2	1	1/2	
		5	1	1	0	

The probability distribution function of the test statistic is $\mathbb{P}(D = 1/3) = 0.1$, $\mathbb{P}(D = 1/2) = 0.3$, $\mathbb{P}(D = 2/3) = 0.4$, and $\mathbb{P}(D = 1) = 0.2$. The cumulative probabilities are $\mathbb{P}(D \leq 1/3) = 0.1$, $\mathbb{P}(D \leq 1/2) = 0.4$, $\mathbb{P}(D \leq 2/3) = 0.8$, and $\mathbb{P}(D \leq 1) = 1$. For larger sample sizes, these tail probabilities become smaller than 0.05 or 0.01, yielding critical values.

1.5.3 SAS Implementation

The default output of the procedure NPAR1WAY includes the results of the Kolmogorov-Smirnov test. The syntax is

```
PROC NPAR1WAY DATA=data_name;
      CLASS sample_name;
         VAR variable_name;
      EXACT;
RUN;
```

• The EXACT statement must appear for the exact test to be performed.
• The output contains also an asymptotic P-value which is computed according to the formula

$$P(D > d) = 2 \sum_{i=1}^{\infty} (-1)^{i-1} \exp \left\{ - \frac{2i^2 d^2 n_1 n_2}{n_1 + n_2} \right\}.$$

1.5.4   Examples

Example 1.13 In Examples 1.8 and 1.11 we concluded that the distributions of the time to full weight bearing do not differ in either location or scale parameters for the control and treatment groups. The natural extension is to test the alternative hypothesis $H_1 : F_{Tx}(t) > F_{Cx}(t)$ for some t, stemming from the fact that, assuming efficacy of the bioimplant, small observations would be more likely in the Tx group than in the Cx group.

The values of the empirical distribution functions and their difference are given in the following table.

Observation	$\hat{F}_{Tx}$	$\hat{F}_{Cx}$	$\hat{F}_{Tx} - \hat{F}_{Cx}$
3	1/6	0	1/6
4	2/6 = 1/3	3/9 = 1/3	0
5	1/3	4/9	−1/9
6	4/6 = 2/3	5/9	1/9
7	2/3	6/9 = 2/3	0
8	6/6 = 1	7/9	2/9
10	1	8/9	1/9
12	1	1	0

The test statistic $D^+ = \max(\hat{F}_{Tx} - \hat{F}_{Cx}) = 2/9$. From Table A.4, the 5% one-sided critical value for the sample sizes of 6 and 9 is 5/9. The test statistic is smaller than the critical value, indicating that the null hypothesis should not be rejected. We conclude that the distribution functions are the same, and hence the drug is not effective.

Running SAS yields the same conclusion. The code is

```
proc npar1way data=bioimplant;
   class group;
      var timeFWB;
   exact;
run;
```

The relevant output is

```
 Kolmogorov-Smirnov Two-Sample Test
 D+ = max (F1 - F2)           0.2222
 Exact   Pr >= D-             0.5385
```

The P-value is much larger than 0.05. Hence, the above conclusion is valid.   □

Example 1.14 A new anti-cancer drug is tested in acute leukemia patients. The readings of the white blood cell counts (in 10^9 cells per liter) are available for eight patients in the treatment group (Tx group) and six patients in the control group (Cx group). The data are:

Tx group	12.33	10.44	12.72	13.13	13.50	16.82	17.60	14.37
Cx group	16.45	18.63	13.12	18.94	19.34	22.50		

In this setting, the Cx group has smaller size than the Tx group, and thus we treat it as the first sample. To test whether the distribution function for the Tx group favors lower readings, we conduct a lower-tailed Kolmogorov-Smirnov test with the alternative hypothesis $H_1 : F_{Cx}(t) < F_{Tx}(t)$ for some t. Indeed, if smaller values are recorded in the Tx group than in the Cx group, then F_{Tx} quickly climbs up and thus exceeds F_{Cx}. The following calculations must be done before the test statistic D^- is computed:

Observation	$\hat{F}_{Cx}$	$\hat{F}_{Tx}$	$\hat{F}_{Tx} - \hat{F}_{Cx}$
10.44	0	1/8	$1/8 = 6/48$
12.33	0	2/8	$2/8 = 12/48$
12.72	0	3/8	$3/8 = 18/48$
13.12	1/6	3/8	10/48
13.13	1/6	4/8	16/48
13.50	1/6	5/8	22/48
14.37	1/6	6/8	28/48
16.45	2/6	6/8	20/48
16.82	2/6	7/8	26/48
17.60	2/6	1	$4/6 = 32/48$
18.63	3/6	1	$3/6 = 24/48$
18.94	4/6	1	$2/6 = 16/48$
19.34	5/6	1	$1/6 = 8/48$
22.50	1	1	0

The value of the test statistic is $D^- = \max(F_{Tx} - F_{Cx}) = 32/48$. As seen in Table A.4, for $n_1 = 6$, $n_2 = 8$, and a one-sided alternative, the critical value for $\alpha = 0.05$ is $7/12 = 28/48$, and for $\alpha = 0.01$, it is $3/4 = 36/48$. Thus, the null hypothesis is rejected at the 5% significance level, and the conclusion is drawn that there is enough evidence in the data to suggest that the new drug is effective. At the 1% level of significance, however, the null hypothesis is not rejected, and the conclusion is opposite to that drawn at the 5% level.

To run SAS for these data, use the code

```
data leukemia;
    input group $ WBCcount @@; /*WBCcount=white blood cell count */
datalines;
Cx 16.45 Cx 18.63 Cx 13.12 Cx 18.94
Cx 19.34 Cx 22.50 Tx 12.33 Tx 10.44
Tx 12.72 Tx 13.13 Tx 13.50 Tx 16.82
Tx 17.60 Tx 14.37
;
```

```
proc npar1way data=leukemia;
   class group;
      var WBCcount;
   exact;
run;
```

The test statistic and P-value are given as

```
Kolmogorov-Smirnov Two-Sample Test
D- = max (F2 - F1)          0.6667
Exact   Pr >= D-            0.0303
```

Note that the P-value is smaller than 0.05 but larger than 0.01, supporting our conclusions stated above. □

Example 1.15 The analysis done in Examples 1.9 and 1.12 led to the conclusion that there are no significant differences either between the location parameters or between the scale parameters of the self-efficacy distributions for men and women. At this point, we might want to test whether the probability distributions of self-efficacy for men and women differ at all. We test the two-sided alternative $H_1 : F_{men}(t) \neq F_{women}(t)$ for some t. The empirical distributions and their absolute difference are presented in the following table:

Observation	$\hat{F}_{men}$	$\hat{F}_{women}$	$\lvert \hat{F}_{men} - \hat{F}_{women} \rvert$
3	$2/6 = 1/3$	0	$1/3 = 0.3333$
4	$5/6$	$4/7$	$11/42 = 0.2619$
5	1	1	0

The test statistic is $D = \max \lvert \hat{F}_{men} - \hat{F}_{women} \rvert = 0.3333$. From Table A.4, the 0.05-critical value for the two-tailed test with samples of size 6 and 7 is $4/7 = 0.5714$. The observed test statistic is less than the critical value, meaning that the null hypothesis cannot be rejected. The conclusion is that the distributions of self-efficacy for men and women and identical.

The same conclusion is reached after running SAS. The complete code is

```
data elderly_study;
      input gender $ self_efficacy @@;
   datalines;
M 4 M 5 M 4 M 3 M 4 M 3
F 5 F 5 F 4 F 4 F 4 F 5 F 4
;

proc npar1way data=elderly_study;
   class gender;
```

```
        var self_efficacy;
    exact;
run;
```

The test statistic and its P-value are given as

```
Kolmogorov-Smirnov Two-Sample Test
D = max |F1 - F2|        0.3333
Exact   Pr >= D          0.4207
```

Our conclusion is supported by the large P-value.   □

Exercises for Chapter 1

Exercise 1.1 A fitness nutrition coach wanted to see for himself whether a certain muscle building food supplement was effective. He conducted a study for which he randomly selected ten trainees. At the start of the study, he recorded the total circumference of both arms for each participant. Then for two weeks the participants took the supplement and continued their regular workouts. The total circumference of both arms was remeasured at the end of the two weeks. The results (in inches) are summarized in the chart below.

Start	Two Weeks
22.125	23.375
23.500	23.125
23.500	24.750
25.875	25.750
26.375	26.750
21.375	22.625
28.875	29.000
24.625	24.000
23.125	24.125
25.000	27.250

Test the claim using the sign test. Show all steps. State your conclusion clearly. Use SAS to verify the results.

Exercise 1.2 A research team of clinical mental health counselors conducted a study of a therapy intervention which goal was to reduce post-traumatic stress in low-income ethnic minority individuals. They suspected that gender and age might play a significant role in how effective the therapy is, so they controlled for these cofactors by matching individuals on gender, age, and the initial score on a standard mental health questionnaire. They randomly assigned individuals to the treatment and wait-list control groups within each matched pair. After the treated individual had completed the therapy, both individuals retook the questionnaire. The increase in scores was recorded. Higher scores indicated a better mental fit. The results were as follows:

Pair	Intervention	Control
1	36	17
2	22	15
3	10	−8
4	12	−11
5	28	14
6	12	20
7	23	24
8	6	6

Carry out the sign test. Is the therapy effective? Do the testing by hand and in SAS.

Exercise 1.3 A market research specialist is investigating the behavior of a stock price for a large stock-issuing company. She records for two consecutive weeks the closing price per share for the trading day (in US$). She is interested in determining whether there is a shift in the weekly distribution of prices. Since prices may vary by day of the week, she conducts a matched paired test. The data are

Day of the Week	Stock Price Week 1	Week 2
Mon	405.65	403.02
Tue	400.51	399.49
Wed	408.25	396.10
Thu	401.34	403.59
Fri	409.09	405.68

Perform a sign test for these data. Draw conclusion. Do the analysis by hand and in SAS.

Exercise 1.4 Apply the Wilcoxon signed-rank test to the data in Exercise 1.1. Do the sign and signed-rank tests produce dissimilar results? Use $\alpha = 0.05$. Do the computations by hand and in SAS.

Exercise 1.5 Use the data from Exercise 1.2 to conduct the Wilcoxon signed-rank test. Does the conclusion coincide with the one of the sign test? Use $\alpha = 0.05$. Verify the results in SAS.

Exercise 1.6 Conduct the Wilcoxon signed-rank test for the data in Exercise 1.3. Can a meaningful conclusion be drawn when doing the calculations by hand? Why or why not? What conclusion is drawn from SAS output?

Exercise 1.7 Knee arthroplasty is a total knee replacement surgery performed for severe degenerative diseases of the knee joint. A minimally invasive arthroplasty involves making a smaller incision, and thus patients who undergo this surgical procedure may expect shorter hospital stay and quicker recovery. To test validity of this claim, a prospective clinical trial was conducted. Ten patients underwent the minimally invasive procedure (Tx group), whereas seven received the traditional arthroplasty (Cx group). The number of days of hospital stay was recorded for each patient. The data follow.

Tx group	5	4	6	4	3	4	4	3	5	5
Cx group	7	8	12	10	8	9	10			

Conduct the Wilcoxon rank-sum test on these data. State the hypotheses. Do the data support the claim? Carry out the analysis by hand and in SAS.

Exercise 1.8 A group of health education researchers is interested in testing the difference in HIV-related knowledge between adolescents and adults. They suspect that adults have a better knowledge. They randomly choose 12 adolescents and 12 adults from the same community and administer a standard HIV knowledge questionnaire that consists of 18 true or false questions. Only eight adolescents and ten adults answer all the questions. The number of correct responses is recorded for each individual who completes the questionnaire. The data are

Adolescents	15	8	8	10	6	9	7	8		
Adults	13	17	10	12	13	17	15	17	17	19

Carry out a one-tailed Wilcoxon rank-sum test at the 1% significance level. Verify your conclusion by running SAS.

Exercise 1.9 A professor of management studies social network among the students who are taking his business course. Since the program is small, he expects that all students know each other. He asks each student to identify how many students in the class he or she likes and respects. He then gathers the information for all students in a chart stratifying by gender. He suspects that gender doesn't play a significant role in this case. The data are

Females	10	7	11	8	5	12	13			
Males	7	6	8	5	3	6	7	6	3	2

Perform the Wilcoxon rank-sum test at $\alpha = 0.01$. Specify the hypotheses, test statistic and critical value(s). Draw your conclusion. Also carry out the analysis in SAS.

Exercise 1.10 A Master's student in kinesiology is studying a sociological aspect of physical activities. She surveys two groups of women, full-time working moms and stay-at-home moms. She collects measurements on the number of hours spent exercising last week. The data are

Working moms	1	4	14	7	11	1	8	10
Stay-at-home moms	1	3	5	5	4	3	4	5

(a) Verify that the Wilcoxon rank-sum test reveals that there is no difference in the location parameters of the underlying distributions in the two groups. Do it by hand

and in SAS.

(b) Conduct the Ansari-Bradley test to check whether the dispersion of the measured variable differs by group. Draw conclusion. Use a 0.05 level of significance. Verify your results in SAS.

(c) Conduct the Ansari-Bradley test to check whether the dispersion of the measured variable is larger for working moms. Use $\alpha = 0.01$ and 0.05. Verify your results in SAS.

Exercise 1.11 An orthopedic surgeon performs rotator cuff repair surgeries in patients with acute shoulder pain. During a follow-up visit, he uses two rating scales to evaluate the level of pain, activities of daily living (sleep, work, recreation/sport), and range of motion of the repaired shoulders. The UCLA shoulder score rating scale has a maximum of 35 points, whereas the Constant shoulder score is a percentage of a total of 100 points.[1] The surgeon normalizes the data and organizes them in the form of a table.

Normalized UCLA Score	27	37	40	63	31	81	63	57	90	94
Constant Shoulder Score	56	78	60	55	67	68	64			

There are more questions that have to be answered when calculating the Constant shoulder score. Based on this fact, the surgeon claims that the Constant shoulder scale calculates scores with a better precision. Verify this claim by performing the Ansari-Bradley test. Use the 0.05 significance level. Do computations by hand and using SAS. Hint: First perform the Wilcoxon rank-sum test to show that there is no difference in the location parameters and the Ansari-Bradley test is applicable.

Exercise 1.12 Performance of a new adhesion barrier sealant is assessed in a post-marketing approval study. The product is a liquid gel that is sprayed on organs to help prevent post-operative adhesions (scar tissue). Adhesions are detected during a follow-up laparoscopic surgical procedure where only a small incision in the abdomen is made. A score is assigned to each patient based on the number of adhesions, their severity and extent. The data are available for five patients treated with the sealant (Tx group) and five non-treated patients (Cx group). A smaller score indicates a better clinical success.

Tx group	2.1	1.6	3.8	3.2	4.0
Cx group	3.7	7.2	2.8	5.3	8.6

Carry out the Kolmogorov-Smirnov test by hand and in SAS to verify the efficacy of the sealant. Use the significance level of 0.05.

Exercise 1.13 A retrospective study is conducted to assess efficacy of chemotherapy for patients with benign non-operative liver tumors. The reduction from the baseline value in tumor diameter (in cm) at the end of a one-month trial period is available

[1]Constant, C.R. and Murley, A.H. (1987) A clinical method of functional assessment of the shoulder, *Clinical Orthopaedics and Related Research*, **214**, 160-164.

for seven treated patients (Tx group) and nine control patients (Cx group). The data are presented below. The negative numbers mean that the tumor diameter actually increased during the study. Conduct the Kolmogorov-Smirnov test with $\alpha = 0.05$ and 0.01. Do the analysis by hand and in SAS.

Tx group	2.5	2.4	2.1	3.4	4.2	1.1	1.9		
Cx group	-0.9	1.5	2.3	-1.6	-3.4	0.3	2.0	-1.1	1.6

Exercise 1.14 Refer to Exercise 1.11. Check whether the scores on the two shoulder pain rating scales have equal distribution functions. Use the Kolmogorov-Smirnov test with $\alpha = 0.05$. Support your findings by running SAS.

Chapter 2

Hypotheses Testing for Several Samples

2.1 Friedman Rank Test for Location Parameter for Several Dependent Samples

When means of more than two dependent samples are compared, the parametric approach calls for the repeated measures analysis of variance. A reasonable non-parametric technique in this case is the Friedman rank test for location parameters. This test was introduced in 1937 by an American economist and statistician Milton Friedman (1912-2006).[1]

2.1.1 Testing Procedure

Suppose k measurements of the same variable are taken on n individuals repeatedly over time or across certain conditions. These k sets of measurements are typically referred to as *dependent samples*, because the same individuals constitute each sample. The null hypothesis states that all k location parameters are equal,

$$H_0 : \theta_1 = \theta_2 = \cdots = \theta_k,$$

and is tested against the alternative hypothesis asserting that not all k location parameters are the same,

$$H_1 : \theta_i \neq \theta_j \text{ for some } i \neq j, i, j = 1, ..., k.$$

To compute the *Friedman rank test* statistic, we first assign ranks to observations within each individual in such a way that the smallest value gets the rank of 1. If two or more observations are tied, assign each of them the same rank which is the mean of the ranks that would have been assigned to these values if they were not tied. Denote by $r_{ij}, i = 1, \ldots, n, j = 1, \ldots, k$, the rank for the measurement on the i-th individual in the j-th sample (for example, on the j-th occasion or for the j-th condition). Let $R_j = \sum_{i=1}^{n} r_{ij}, j = 1, \ldots, k$, be the sum of the ranks of all measurements in the j-th sample. In case there are no ties, the test statistic Q is derived as

$$Q = \frac{12}{nk(k+1)} \sum_{j=1}^{k} R_j^2 - 3n(k+1). \tag{2.1}$$

---

[1]Friedman, M. (1937) The use of ranks to avoid the assumption of normality implicit in the analysis of variance, *Journal of the American Statistical Association*, **32**, 675-701.

In the presence of ties, the test statistic is determined by the formula

$$Q = \frac{n(k-1)\left[(1/n)\sum_{j=1}^{k} R_j^2 - nk(k+1)^2/4\right]}{\sum_{i=1}^{n}\sum_{j=1}^{k} r_{ij}^2 - nk(k+1)^2/4}. \qquad (2.2)$$

Note that (2.1) follows from (2.2) when all ranks are different. Indeed, the sum of squares of k distinct ranks for an ith individual is computed explicitly as

$$\sum_{j=1}^{k} r_{ij}^2 = n\left[1^2 + 2^2 + \cdots + k^2\right] = nk(k+1)(2k+1)/6.$$

Now simple algebraic manipulations yield the result:

$$Q = \frac{n(k-1)\left[\sum_{j=1}^{k} R_j^2/n - nk(k+1)^2/4\right]}{nk(k+1)(2k+1)/6 - nk(k+1)^2/4}$$

$$= \frac{n(k-1)\left[\sum_{j=1}^{k} R_j^2/n - nk(k+1)^2/4\right]}{nk(k+1)(k-1)/12} = \frac{12}{nk(k+1)} \sum_{j=1}^{k} R_j^2 - 3n(k+1).$$

Next, the test statistic Q is compared to an appropriate critical value found in Table A.5 in Appendix A. The tabulated critical values depend on k, n, and the significance level α. If the test statistic is less than the critical value, the null hypothesis is not rejected, and the conclusion is that the location parameters in all k samples are the same. If Q is larger than or equal to the critical value, H_0 is rejected in favor of H_1, and the conclusion is that not all location parameters are equal.

Note that the alternative for the multi-sample comparison doesn't provide information on which location parameters differ and in which direction. The Wilcoxon signed-rank test may be used further to do the pairwise comparison of the location parameters.

2.1.2   SAS Implementation

To run the Friedman rank test in SAS, the data must be presented in the *long form*, that is, in the form individual-sample-response with k input lines for every individual, a total of nk lines. The generic code then is:

```
PROC SORT data=data_name;
      BY individual_name;
RUN;

PROC RANK DATA=data_name OUT=outdata_name;
      VAR response_name;
         BY individual_name;
            RANKS rank_name;
```

```
RUN;

PROC FREQ data=outdata_name;
    TABLE individual_name*sample_name*rank_name/ NOPRINT CMH;
RUN;
```

- The SORT procedure sorts data in preparation for computing ranks within each individual.
- The RANK procedure assigns ranks, which are output in a different data set along with the original variables.
- The FREQ procedure is called to create a three-way table which, thanks to the NOPRINT option, is not printed.
- The Friedman rank test statistic is requested by adding the option CMH. It stands for Cochran-Mantel-Haenszel test, for which the Friedman rank test is a special case.
- The Friedman rank test statistic is <u>number two</u> in the output window among the three statistics computed by SAS.
- The P-value is calculated based on the chi-square distribution with $k - 1$ degrees of freedom, which is a large-sample asymptotic distribution of the Friedman test statistic.

2.1.3 Examples

Example 2.1 A psychology student working on his Ph.D. thesis has written a computer software that creates choice reaction time tasks. He would like to test if a difference in response is present across three task conditions: pressing button with both hands, with only right hand, or with only left hand. He randomly selects six people for his study. For each person, he randomly chooses the order in which task conditions are applied. The multiple readings of their reaction times (in seconds) and the assigned ranks are recorded below.

Individual	Both Hands	Right Hand	Left Hand
1	1.0	0.6	0.7
Rank	3	1	2
2	0.4	0.5	0.6
Rank	1	2	3
3	0.2	0.5	0.4
Rank	1	3	2
4	0.3	0.2	0.5
Rank	2	1	3
5	0.4	0.3	0.5
Rank	2	1	3
6	0.1	0.2	0.3
Rank	1	2	3
Rank Total	10	10	16

In the absence of tied observations, we use formula (2.1) for derivation of the Friedman rank test statistic. We have that $k = 3$, $n = 6$, $R_1 = 10$, $R_2 = 10$, and $R_3 = 16$. Therefore,

$$Q = \frac{12}{nk(k+1)} \sum_{j=1}^{k} R_j^2 - 3n(k+1)$$

$$= \frac{12}{(6)(3)(3+1)} \left(10^2 + 10^2 + 16^2\right) - (3)(6)(3+1) = 4.$$

From Table A.5, the critical value for $k = 3$, $n = 6$, and $\alpha = 0.05$ is 7. The test statistic is less than the critical value, hence, the null hypothesis $H_0 : \theta_{both} = \theta_{right} = \theta_{left}$ cannot be rejected. The conclusion is that there is no difference across the conditions.

SAS is applied to verify the calculations and the conclusion. The code is

```
data reaction_time;
     input individual hand $ time @@;
   datalines;
1 both 1.0 1 right 0.6 1 left 0.7
2 both 0.4 2 right 0.5 2 left 0.6
3 both 0.2 3 right 0.5 3 left 0.4
4 both 0.3 4 right 0.2 4 left 0.5
5 both 0.4 5 right 0.3 5 left 0.5
6 both 0.1 6 right 0.2 6 left 0.3
;

proc sort data=reaction_time;
   by individual;
run;

proc rank data=reaction_time out=ranked;
   var time;
      by individual;
         ranks rank;
run;

proc freq data=ranked;
   table individual*hand*rank/noprint cmh;
run;
```

The partial output is

```
Alternative Hypothesis   Value    Prob
Row Mean Scores Differ   4.0000   0.1353
```

The large P-value $= 0.1353$ indicates that the null hypothesis should not be rejected. □

Example 2.2 A manager in a brewery conducts a beer tasting study. Ten participants are randomly recruited among adults who have drunk a specified minimum amount of beer in a recent time period. Each taster is given three brands of beer (A, B, and C). The sequence in which beers are tasted is randomly determined for each person. For each brand, they fill out an opinion questionnaire that rates a beer preference on a nine-point scale. A beer is considered "good" if it is given the preference rating of 6 or higher; and "bad," otherwise. The manager would like to see whether the beer ratings are different. The data and the corresponding ranks are

Beer	Beer Preference Rating		
Taster	Brand A	Brand B	Brand C
1	4	5	8
Rank	1	2	3
2	3	7	6
Rank	1	3	2
3	2	8	8
Rank	1	2.5	2.5
4	5	7	9
Rank	1	2	3
5	1	3	6
Rank	1	2	3
6	4	4	6
Rank	1.5	1.5	3
7	3	6	8
Rank	1	2	3
8	2	5	5
Rank	1	2.5	2.5
9	2	7	9
Rank	1	2	3
10	6	4	5
Rank	3	1	2
Rank Total	12.5	20.5	27.0

Since ties are present, equation (2.2) is used to compute the test statistic. The quantities needed for calculations are $n = 10$, $k = 3$, $R_1 = 12.5$, $R_2 = 20.5$, $R_3 = 27.0$, and $\sum_{i=1}^{10} \sum_{j=1}^{3} r_{ij}^2 = 138.5$. The test statistic is

$$Q = \frac{n(k-1)\left[(1/n)\sum_{j=1}^{k} R_j^2 - nk(k+1)^2/4\right]}{\sum_{i=1}^{n}\sum_{j=1}^{k} r_{ij}^2 - nk(k+1)^2/4}$$

$$= \frac{10(3-1)\left[(1/10)(12.5^2 + 20.5^2 + 27.0^2) - (10)(3)(3+1)^2/4\right]}{138.5 - (10)(3)(3+1)^2/4} = 11.4054.$$

The critical value from Table A.5 for $k = 3, n = 10$, and $\alpha = 0.05$ is 6.2. For $\alpha = 0.01$, it is 9.6. Thus, the null hypothesis $H_0 : \theta_A = \theta_B = \theta_C$ is rejected even at the 0.01 level of significance and the conclusion is that tasted beer brands differ in preference ratings.

To conduct the Friedman rank test in SAS, we use the code

```
data beer;
     input taster brand $ rating @@;
   datalines;
1 A 4 1 B 5 1 C 8 2 A 3 2 B 7 2 C 6 3 A 2 3 B 8 3 C 8
4 A 5 4 B 7 4 C 9 5 A 1 5 B 3 5 C 6 6 A 4 6 B 4 6 C 6
7 A 3 7 B 6 7 C 8 8 A 2 8 B 5 8 C 5 9 A 2 9 B 7 9 C 9
10 A 6 10 B 4 10 C 5
;

proc sort data=beer;
   by taster;
run;

proc rank data=beer out=ranked;
   var rating;
      by taster;
         ranks rank;
run;

proc freq data=ranked;
   table taster*brand*rank/noprint cmh;
run;
```

The output pertained to the problem at hand is

```
Alternative Hypothesis     Value     Prob
Row Mean Scores Differ    11.4054   0.0033
```

The small P-value supports our conclusion that the ratings differ. To see exactly which ratings are not the same, conduct the Wilcoxon signed-rank testing to compare brands A and B, A and C, and B and C. To this end, compute the differences and their ranks as shown in the table below.

Beer	Beer Preference Rating								
Taster	Brand A	Brand B	Brand C	A−B	Rank	A−C	Rank	B−C	Rank
1	4	5	8	−1	1	−4	5.5	−3	7.5
2	3	7	6	−4	7	−3	3.5	1	1.5
3	2	8	8	−6	9	−6	9	0	−
4	5	7	9	−2	3	−4	5.5	−2	4.5
5	1	3	6	−2	3	−5	7.5	−3	7.5
6	4	4	6	0	−	−2	2	−2	4.5
7	3	6	8	−3	5.5	−5	7.5	−2	4.5
8	2	5	5	−3	5.5	−3	3.5	0	−
9	2	7	9	−5	8	−7	10	−2	4.5
10	6	4	5	2	3	1	1	−1	1.5

- To test $H_0 : \theta_A = \theta_B$ against $H_1 : \theta_A \neq \theta_B$, compare the test statistic $T = \min(T^+, T^-) = T^+ = 3$ to the two-sided critical value $T_0 = 5$ from Table A.1 that corresponds to $n = 9$ and $\alpha = 0.05$. Thus, the null hypothesis is rejected, and the conclusion is that the ratings differ for brands A and B at the 5% significance level. Note that T_0 for $\alpha = 0.01$ is equal to 1, so at the 0.01 level, the null should not be rejected.

- To test $H_0 : \theta_A = \theta_C$ against $H_1 : \theta_A \neq \theta_C$, compute the test statistic $T = \min(T^+, T^-) = T^+ = 1$, and compare it to a two-tailed critical value T_0 for $n = 10$. At $\alpha = 0.01$, $T_0 = 3$. Therefore, the null hypothesis is rejected even at the 1% significance level, and conclusion is that the ratings for brands A and C differ.

- To test $H_0 : \theta_B = \theta_C$ against $H_1 : \theta_B \neq \theta_C$, note that the test statistic $T_0 = \min(T^+, T^-) = T^+ = 1.5$ is between the two-sided critical values $T_0 = 0$ for $\alpha = 0.01$ (and $n = 8$), and $T_0 = 3$ for $\alpha = 0.05$. Hence, the null is rejected at the 5% but not 1% significance level, and the conclusion is that the brands B and C differ in their ratings at the 5% but not at 1%.

To verify the findings in SAS, we run the following lines of code:

```
data tasting;
   input A B C;
      diff_AB=A-B;
         diff_AC=A-C;
            diff_BC=B-C;
cards;
4 5 8
3 7 6
2 8 8
5 7 9
1 3 6
```

```
4 4 6
3 6 8
2 5 5
2 7 9
6 4 5
;

proc univariate data=tasting;
   var diff_AB diff_AC diff_BC;
run;
```

The relevant output is given as

• Comparison between brands A and B: $S = T^+ - n(n+1)/4 = 3 - 9(9+1)/4 = 3 - 22.5 = -19.5$.

```
Test                 -Statistic-      ----p Value----
Signed Rank       S      -19.5      Pr >= |S|   0.0195
```

• Comparison between brands A and C: $S = T^+ - n(n+1)/4 = 1 - 10(10+1)/4 = 1 - 27.5 = -26.5$.

```
Test                 -Statistic-      ----p Value----
Signed Rank       S      -26.5      Pr >= |S|   0.0039
```

• Comparison between brands B and C: $S = T^+ - n(n+1)/4 = 1.5 - 8(8+1)/4 = 1.5 - 18 = -16.5$.

```
Test                 -Statistic-      ----p Value----
Signed Rank       S      -16.5      Pr >= |S|   0.0234
```

The conclusions drawn from this output are equivalent to the earlier ones. According to these P-values, brands A and B differ at the 5% but not the 1% significance level (P-value $= 0.0195$); brands A and C differ at the 0.01 level (P-value $= 0.0039$); and brands B and C differ at the level of 0.05 but not 0.01 (P-value $= 0.0234$).   □

2.2   Kruskal-Wallis H-Test for Location Parameter for Several Independent Samples

If data are coming from several independent samples and the goal is to test equality versus non-equality of the respective location parameters, then one should conduct a one-way analysis of variance (ANOVA). A nonparametric equivalent of the ANOVA approach is the *Kruskal-Wallis H-test* named after two American statisticians who formulated it in 1952, William H. Kruskal (1919-2005) and W. Allen Wallis (1912-1998).[1]

---

[1] Kruskal, W.H. and Wallis, W.A. (1952) Use of ranks in one-criterion variance analysis, *Journal of the American Statistical Association*, **47**, 583-621.

2.2.1 Testing Procedure

Suppose measurements for k independent samples of respective sizes $n_1, \ldots, n_k$ are available. The hypotheses of interest are

$$H_0 : \theta_1 = \theta_2 = \cdots = \theta_k$$

and

$$H_1 : \theta_i \neq \theta_j \text{ for some } i \neq j, i, j = 1, \ldots, k.$$

To compute the Kruskal-Wallis test statistic, we first pool all observations together and assign ranks in increasing order. Tied values are given ranks equal to the average of the ranks that would have been assigned were they not tied. Denote by $R_i, i = 1, \ldots, k$, the sum of the ranks in the i-th sample. Let $N = n_1 + \cdots + n_k$ be the total number of observations. If no ties exist, the H statistic is determined according to the formula

$$H = \frac{12}{N(N+1)} \sum_{i=1}^{k} \frac{R_i^2}{n_i} - 3(N+1). \tag{2.3}$$

Suppose now that the ranking procedure results in m sets of ties. Let $T_i, i = 1, \ldots, m$, denote the number of ties in the i-th set. Then the H statistic is computed as

$$H = \left[\frac{12}{N(N+1)} \sum_{i=1}^{k} \frac{R_i^2}{n_i} - 3(N+1) \right] \bigg/ \left[1 - \frac{\sum_{i=1}^{m}(T_i^3 - T_i)}{N^3 - N} \right]. \tag{2.4}$$

Note that if no ties occur, all T_i's are equal to one, and (2.4) is simplified to the form given in (2.3).

Once the test statistic is computed, it is compared to the appropriate critical value from Table A.6 in Appendix A. If H is larger than or equal to the critical value, the null hypothesis is rejected, and the conclusion is that not all k location parameters are the same. To check which location parameters differ from each other, we conduct the Wilcoxon rank-sum tests for pairs of samples.

2.2.2 SAS Implementation

The code lines requesting the Kruskal-Wallis test in SAS are identical to the ones for the Wilcoxon rank-sum test, introduced in Subsection 1.3.3. For convenience we repeat the syntax here.

```
PROC NPAR1WAY DATA=data_name WILCOXON;
        CLASS sample_name;
            VAR variable_name;
        EXACT;
RUN;
```

• Along with the exact P-value, SAS outputs an asymptotic one based on the chi-square distribution with $k-1$ degrees of freedom, which is a large-sample asymptotic distribution of the Kruskal-Wallis test statistic.
• In case a significant difference between the samples is detected, do a post-hoc pairwise comparison using the Wilcoxon rank-sum test. To select any two samples from the data set, use the WHERE clause:
WHERE (*sample_name*='sample_1' OR *sample_name*='sample_2');

2.2.3 Examples

Example 2.3 An Archery 101 instructor held a tournament among her three students Monica, Bob, and Jeff. The target scoring system was 10 points for the middle X circle and 5 through 1 points for the bigger circles. Each student made 10 shots. Some of these shots missed the target. Based on the total score, Jeff was the winner. The instructor, however, decided to determine the winner statistically. She presented ordered data and the respective ranks in a table, with the last row containing the score and rank totals.

Monica		Bob		Jeff	
Score	Rank	Score	Rank	Score	Rank
3	6.5	2	4	1	1.5
4	10.5	2	4	1	1.5
4	10.5	3	6.5	2	4
5	16	4	10.5	4	10.5
5	16	4	10.5	4	10.5
5	16	5	16	5	16
10	22	10	22	10	22
10	22	10	22	10	22
				10	22
46	119.5	40	95.5	47	110

The alternative hypothesis is that there is a difference in location parameters of the score distributions for these three students. The null hypothesis is that the location parameters are all the same. The ranking procedure results in several sets of tied observations. Thus we use the expression (2.4) to determine the test statistic. We have that $n_1 = n_2 = 8$, $n_3 = 9$, $N = 8+8+9 = 25$, $R_1 = 119.5$, $R_2 = 95.5$, and $R_3 = 110$. Also, the number of 1's is $T_1 = 2$, the number of 2's is $T_2 = 3$, the number of 3's is $T_3 = 2$, the number of 4's is $T_4 = 6$, the number of 5's is $T_5 = 5$, and the number of 10's is $T_6 = 7$. The denominator in (2.4) is equal to $1 - \dfrac{\sum_{i=1}^{m}(T_i^3 - T_i)}{N^3 - N} = 1 - $
$\dfrac{(2^3 - 2) + (3^3 - 3) + (2^3 - 2) + (6^3 - 6) + (5^3 - 5) + (7^3 - 7)}{25^3 - 25} = 0.9550$, and hence, the H-statistic is computed as

$$H = \left[\frac{12}{N(N+1)} \sum_{i=1}^{k} \frac{R_i^2}{n_i} - 3(N+1) \right] \Big/ \left[1 - \frac{\sum_{i=1}^{m}(T_i^3 - T_i)}{N^3 - N} \right]$$

$$= \left[\frac{12\left(119.5^2/8 + 95.5^2/8 + 110^2/9\right)}{(25)(25+1)} - 3(25+1)\right]/0.9550 = 0.8604.$$

From Table A.6, the critical value for sample sizes 9, 8, and 8, and $\alpha = 0.05$ is 5.810, which is above the test statistic. Therefore, we fail to reject the null hypothesis and draw the conclusion that, statistically speaking, there is no clear winner in this tournament.

The same value of the test statistic can be obtained in SAS. The code is

```
data archery;
     input name $ score @@;
  datalines;
Monica 3 Monica 4 Monica 4 Monica 5 Monica 5 Monica 5 Monica 10
Monica 10 Bob 2 Bob 2 Bob 3 Bob 4 Bob 4 Bob 5 Bob 10 Bob 10
Jeff 1 Jeff 1 Jeff 2 Jeff 4 Jeff 4 Jeff 5 Jeff 10 Jeff 10 Jeff 10
;

proc npar1way data=archery wilcoxon;
  class name;
     var score;
  exact;
run;
```

The output pertained to the Kruskal-Wallis test is given as

```
            Kruskal-Wallis Test
  Chi-Square                 0.8604
  Exact        Pr >= Chi-Square  0.6650
```

Note that the P-value is much larger than 0.05, validating the conclusion. □

Example 2.4 A big fitness center is interested in determining whether aerobics, pilates, step, and cardio classes differ in effectiveness of weight reduction. Twenty overweight women are randomly allocated to the four classes, five women in each class. After an eight-week session, the percentage of excess body weight loss (%EWL) was recorded for each participant. During the course of the study, three women dropped out of the program. The data and rank values are summarized below.

Aerobics	Rank	Pilates	Rank	Step	Rank	Cardio	Rank
6.7	1	10.5	6	13.0	11	19.0	16
7.7	2	12.8	10	11.2	7	15.3	14
10.0	5	13.1	12	11.8	9	17.5	15
9.4	4	13.4	13	11.6	8	22.4	17
9.1	3						
Total	15		41		35		62

We will test the alternative hypothesis that the location parameters of the %EWL distribution are not all the same for these four classes. First we use the Kruskal-Wallis procedure to test equality of underlying location parameters versus non-equality. We have $n_1 = 5$, $n_2 = n_3 = n_4 = 4$, $N = 5 + 4 + 4 + 4 = 17$, $R_1 = 15$, $R_2 = 41$, $R_3 = 35$, and $R_4 = 62$. Since the ranks have no tied values, formula (2.3) is used to compute the H-statistic,

$$H = \frac{12}{N(N+1)} \sum_{i=1}^{k} \frac{R_i^2}{n_i} - 3(N+1)$$

$$= \frac{12\left(15^2/5 + 41^2/4 + 35^2/4 + 62^2/4\right)}{(17)(17+1)} - 3(17+1) = 13.9412.$$

The 1% critical value corresponding to the sample sizes 5, 4, 4, and 4 is 9.392. The null hypothesis must be rejected, and we reach the conclusion that the classes differ in effectiveness of weight reduction at the 1% significance level.

To compute the test statistic and the P-value, we submit the following code to SAS:

```
data weight_loss;
      input fitness_class $ percEWL @@;
   datalines;
aerobics 6.7 aerobics 7.7 aerobics 10.0 aerobics 9.4 aerobics 9.1
pilates 10.5 pilates 12.8 pilates 13.1 pilates 13.4
step 13.0 step 11.2 step 11.8 step 11.6
cardio 19.0 cardio 15.3 cardio 17.5 cardio 22.4
;

proc npar1way data=weight_loss wilcoxon;
   class fitness_class;
      var percEWL;
   exact;
run;
```

The result is

```
            Kruskal-Wallis Test
  Chi-Square                      13.9412
  Exact        Pr >= Chi-Square   5.597E-06
```

The P-value is much smaller than 0.01, thus supporting our conclusion.

The next step is to perform pairwise comparisons by running the Wilcoxon rank-sum tests. We start with ordering the ranks (15, 35, 42, and 62) and conducting two-sided tests comparing only neighboring samples, that is, we make comparisons between aerobics and step classes, step and pilates classes, and pilates and cardio classes.

- To test whether step and aerobics classes differ, $H_1 : \theta_{step} \neq \theta_{aerobics}$, we rank the data as follows:

Step	Rank	Aerobics	Rank
13.0	9	6.7	1
11.2	6	7.7	2
11.8	8	10.0	5
11.6	7	9.4	4
		9.1	3
Total	30		15

The test statistic $W = 30$ is the sum of the ranks in the smaller sample. According to Table A.2, the upper-tailed critical value for $\alpha = 0.05$, $n_1 = 4$, and $n_2 = 5$ is $W_U = 28$. Since $W > W_U$, the null is rejected and the conclusion is that step and aerobics classes differ in effectiveness at the 5% significance level. The critical value for $\alpha = 0.01$ is not listed in the table, meaning that at the 1% level the null hypothesis would not be rejected.

- To test whether pilates and step classes differ, $H_1 : \theta_{pilates} \neq \theta_{step}$, we compute the ranks as given in the table:

Pilates	Rank	Step	Rank
10.5	1	13.0	6
12.8	5	11.2	2
13.1	7	11.8	4
13.4	8	11.6	3
Total	21		15

The test statistic $W = 21$ is the sum of the ranks in the first sample (since samples sizes are equal). From Table A.2, the upper-tailed critical value for $\alpha = 0.05$ and $n_1 = n_2 = 4$ is $W_U = 26$ and the lower-tailed one is $W_L = 10$. Since $W_L < W < W_U$, the null hypothesis must not be rejected leading to the conclusion that pilates and step classes are equally effective tools at the 5% level (and, therefore, also at the 1% level).

- To test whether cardio class differs from pilates, $H_1 : \theta_{cardio} \neq \theta_{pilates}$, the data for these two samples are ranked as listed below.

Cardio	Rank	Pilates	Rank
19.0	7	10.5	1
15.3	5	12.8	2
17.5	6	13.1	3
22.4	8	13.4	4
Total	26		10

The test statistic $W = 26$ is the sum of the ranks in the first sample (since samples sizes are equal). From Table A.2, the upper-tailed critical value for $\alpha = 0.05$ and

$n_1 = n_2 = 4$ is $W_U = 26 = W$. Hence, we reject the null hypothesis and conclude that cardio class and pilates differ in effectiveness, if stating at the 5% significance level. For $\alpha = 0.01$, the critical value W_U is not given in the table, meaning that H_0 must not be rejected. Thus, at the 0.01 level, pilates and cardio do not differ.

• Since we concluded that at the 1% significance level there is no difference between pilates and step, and also between cardio and pilates, we still must carry out a two-sided test comparing cardio and step classes. But this test is trivial since for $n_1 = n_2 = 4$, Table A.2 doesn't provide the critical values and the null is never rejected.

The SAS code for the multiple comparisons is as follows:

```
proc npar1way data=weight_loss wilcoxon;
   class fitness_class;
      var percEWL;
         exact;
      where (fitness_class='step' or fitness_class='aerobics');
run;

proc npar1way data=weight_loss wilcoxon;
   class fitness_class;
      var percEWL;
         exact;
      where (fitness_class='pilates' or fitness_class='step');
run;

proc npar1way data=weight_loss wilcoxon;
   class fitness_class;
      var percEWL;
         exact;
      where (fitness_class='cardio' or fitness_class='pilates');
run;

proc npar1way data=weight_loss wilcoxon;
   class fitness_class;
      var percEWL;
         exact;
      where (fitness_class='cardio' or fitness_class='step');
run;
```

Submitting this code produces the output below.

• To compare step and aerobics:

```
Wilcoxon Two-Sample Test
Statistic (S) 30.0000
Exact Test
Two-Sided Pr >= |S - Mean| 0.0159
```

• To compare pilates and step:

```
Wilcoxon Two-Sample Test
Statistic (S) 21.0000
Exact Test
Two-Sided Pr >= |S - Mean| 0.4857
```

• To compare cardio and pilates:

```
Wilcoxon Two-Sample Test
Statistic (S) 10.0000
Exact Test
Two-Sided Pr >= |S - Mean| 0.0286
```

• To compare cardio and step:

```
Wilcoxon Two-Sample Test
Statistic (S) 10.0000
Exact Test
Two-Sided Pr >= |S - Mean| 0.0286
```

The same conclusion as above is drawn based on these P-values. □

Exercises for Chapter 2

Exercise 2.1 A high school substitute teacher was appalled by how little high school students read for pleasure. She decided to conduct a *crossover* experiment of the form ABAB where A stands for a time period without intervention and B indicates a time period with intervention. She asked her students to report every month how many books they had read that were not an assigned reading. The first and third months she didn't offer any incentives, but the second and fourth months she offered a $10 gift card to a retail store for a person who read the largest number of books. To assure honest answers, she required a paragraph-long summary of each book read. The data for seven individuals are summarized in the table below. Conduct the Friedman rank test (by hand and in SAS) to investigate whether the number of books read per month differ for these four months. If necessary, conduct the two-sided Wilcoxon sign-rank testing to see which months differ.

Student	Month 1	Month 2	Month 3	Month 4
1	2	4	3	4
2	0	1	3	4
3	4	5	4	7
4	3	3	4	3
5	0	0	1	3
6	4	3	5	5
7	5	5	4	2

Exercise 2.2 A TV company is studying the ways its customers prefer to be contacted. Three methods of contact are under investigation: sending a letter in the mail, calling on the residential phone, or sending a text message to the cell phone. Eight customers are available for the study. They are contacted by the customer service department by all three methods (in randomly chosen order). For every contact they assess how happy they are with the service by estimating the probability of staying with the company. Their responses are summarized in the table below. Carry out the Friedman rank test. Is there a difference between the methods of customer contact? Which methods differ, if any? Use $\alpha = 0.01$. Repeat the analysis using SAS. Hint: To see which contact methods differ, conduct the two-sided Wilcoxon sign-rank tests.

Customer	Letter	Phone	Text
1	0.4	0.3	0.1
2	0.8	0.4	0.3
3	0.5	0.4	0.1
4	0.7	0.6	0.2
5	0.6	0.3	0.2
6	0.6	0.5	0.4
7	0.6	0.4	0.3
8	0.7	0.6	0.2

Exercise 2.3 Ecologists perform a study to compare lead contents in three fish ponds chosen at random nationwide. The fish ponds are tested at six different locations and the lead contents (in ppb – parts of lead per billion parts of water) are recorded. The data are:

Fish Pond	Lead Contents (in ppb)					
A	3	4	4	5	7	8
B	10	11	11	12	15	18
C	4	5	6	6	9	10

Perform the Kruskal-Wallis H-test at the 5% significance level. If needed, conduct the Wilcoxon rank-sum tests to identify which fish ponds are different in lead contents. Use $\alpha = 0.05$. Confirm your findings by running SAS.

Exercise 2.4 A botanist studies the germination of seeds of an exotic flower under four different temperatures. She plants the seeds in her laboratory and records the germination rates. The table below displays the data.

	Temperature		
24°C	28°C	32°C	36°C
88	67	93	86
54	72	82	87
65	76	84	81
55	80	78	73

Carry out the Kruskal-Wallis H-test and, if necessary, the Wilcoxon rank-sum tests for pairwise comparisons. Use $\alpha = 0.05$. Run the appropriate SAS code.

Chapter 3

Tests for Categorical Data

A categorical variable is measured on the *ordinal* scale if the categories have natural ordering. For example, size (XS, S, M, L, XL), health (poor, fair, good, excellent), grades (A, B, C, D, F), education (less than high school, high school graduate, above high school). A variable measured on an ordinal scale is referred to as an *ordinal* variable.

A categorical variable is measured on the *nominal* scale if there is no natural ordering to the categories (they can be treated as names). For instance, opinion (yes, no, don't know), political affiliation (democrat, republican, independent, other), religious affiliation (Protestant, Catholic, other), race (African-American, Latino, Caucasian, Asian, Pacific Islander, other). A nominal scale variable is called a *nominal* variable.

In this chapter we study nonparametric tests for correlation between two ordinal variables, and for association between two nominal variables (or one nominal and one ordinal). The test for correlation may also be carried out when one or both variables are non-normal continuous. We provide a variety of examples of that in the next section.

3.1 Spearman Rank Correlation Coefficient Test

Suppose n pairs of realizations (x_i, y_i) are available for continuous variables X and Y. The Pearson product-moment correlation coefficient may be computed as a measure of direction and strength of linear relationship between X and Y. It is defined as

$$r_p = \frac{\sum_{i=1}^{n}(x_i - \bar{x})(y_i - \bar{y})}{\sqrt{\sum_{i=1}^{n}(x_i - \bar{x})^2}\sqrt{\sum_{i=1}^{n}(y_i - \bar{y})^2}}. \tag{3.1}$$

Its computational formula is

$$r_p = \frac{n\sum_{i=1}^{n}x_i y_i - \sum_{i=1}^{n}x_i \sum_{i=1}^{n}y_i}{\sqrt{n\sum_{i=1}^{n}x_i^2 - (\sum_{i=1}^{n}x_i)^2}\sqrt{n\sum_{i=1}^{n}y_i^2 - (\sum_{i=1}^{n}y_i)^2}}. \tag{3.2}$$

This coefficient was developed by Karl Pearson (1857-1936), an eminent English mathematician and statistician.

To test equality to zero of the Pearson correlation coefficient, an assumption is made that X and Y are approximately normally distributed. In case observations drastically deviate from normal distribution, or at least one of the variables is ordinal, the Spearman rank correlation coefficient is a nonparametric alternative to the Pearson correlation coefficient. The method is attributed to Charles Edward Spearman (1863-1945), an English psychologist who had made many contributions to statistics.[1]

3.1.1 Computation of Spearman Correlation Coefficient

Suppose that n pairs of observations are obtained on variables X and Y. Suppose that either the data are non-normal or at least one of these variables is measured on an ordinal scale. The *Spearman rank correlation coefficient* is defined by the same formula (3.1) as the Pearson product-moment correlation, except that computations are done not on the original values but on their ranks. The ranks are assigned separately to X and Y variables. The assignment procedure is as follows: the smallest value gets a rank of 1, the next smallest, the rank of 2, and so on. The largest value gets the rank of n. If tied observations are encountered, each of them is assigned the average of the ranks that they would receive if they were not tied. With a little abuse of notation, we will denote by x_i and y_i the <u>ranks</u> of the respective observations. In this notation, the Spearman rank correlation coefficient is defined precisely as in (3.1),

$$r_s = \frac{\sum_{i=1}^{n}(x_i - \bar{x})(y_i - \bar{y})}{\sqrt{\sum_{i=1}^{n}(x_i - \bar{x})^2}\sqrt{\sum_{i=1}^{n}(y_i - \bar{y})^2}}.$$

The formula (3.2) may be used for computational purposes. However, since we are dealing with ranks, computations are simplified tremendously.

- If there are no ties, r_s satisfies

$$r_s = 1 - \frac{6\sum_{i=1}^{n}(x_i - y_i)^2}{n^3 - n}. \tag{3.3}$$

Indeed, the terms in the denominator in (3.2) become

$$n\sum_{i=1}^{n}x_i^2 - \left(\sum_{i=1}^{n}x_i\right)^2 = n\sum_{i=1}^{n}y_i^2 - \left(\sum_{i=1}^{n}y_i\right)^2$$

$$= \frac{n^2(n+1)(2n+1)}{6} - \frac{n^2(n+1)^2}{4} = \frac{n^2(n+1)(n-1)}{12} = \frac{n^4 - n^2}{12}.$$

Also, $\sum_{i=1}^{n}x_i = \sum_{i=1}^{n}y_i = n(n+1)/2$. Hence, from (3.2),

$$r_s = \frac{n\sum_{i=1}^{n}x_i y_i - n^2(n+1)^2/4}{(n^4 - n^2)/12} = \frac{12\sum_{i=1}^{n}x_i y_i}{n^3 - n} - \frac{3(n+1)}{n-1}.$$

---

[1]Spearman, C. (1904) The proof and measurement of association between two things, *The American Journal of Psychology*, **15**, 72-101.

On the other hand, (3.3) can be rearranged to give the same result:

$$1 - \frac{6\sum_{i=1}^{n}(x_i - y_i)^2}{n^3 - n} = \frac{12\sum_{i=1}^{n}x_iy_i}{n^3 - n} + 1 - \frac{12\sum_{i=1}^{n}x_i^2}{n^3 - n} = \frac{12\sum_{i=1}^{n}x_iy_i}{n^3 - n} - \frac{3(n+1)}{n-1}.$$

• If ties are present, the Spearman coefficient is computed in the following manner. Let there be m sets of tied ranks for x-observations, and let T_i, $i = 1, \ldots, m$, denote the number of ties in the i-th set. Define $T_x = \sum_{i=1}^{m}(T_i^3 - T_i)$. Similarly, T_y is defined for the tied ranks of y-observations. Then r_s is given by the formula:

$$r_s = \frac{n^3 - n - 6\sum_{i=1}^{n}(x_i - y_i)^2 - (T_x + T_y)/2}{\sqrt{(n^3 - n)^2 - (T_x + T_y)(n^3 - n) + T_xT_y}}. \tag{3.4}$$

Clearly, if no ties occur, $T_x = T_y = 0$ and (3.4) simplifies to become (3.3).

3.1.2   Testing Procedure

The Spearman rank correlation coefficient r_s is used to test statistically whether the variables X and Y are correlated in the population. Denote by ρ_s the unknown theoretical value of the Spearman correlation coefficient in the population. The null hypothesis $H_0 : \rho_s = 0$ indicates that the variables are uncorrelated. For this test three types of alternative hypothesis are distinguished:

• $H_1 : \rho_s > 0$ specifies a positive correlation between X and Y. This null is rejected if the computed value r_s is larger than or equal to the tabulated one-tailed critical value at a pre-specified level of significance (see below). The null hypothesis is not rejected otherwise.

• $H_1 : \rho_s < 0$ claims that the correlation between X and Y is negative. This null is rejected if the computed value r_s is negative and its absolute value is greater than or equal to the appropriate tabulated one-tailed critical value (see below). The null hypothesis is not rejected otherwise.

• $H_1 : \rho_s \neq 0$ states that a non-zero correlation between X and Y exists but its direction is of no importance or is a priori unknown. This null is rejected if the absolute value of r_s is not less than the appropriate tabulated two-tailed critical value (see below). Otherwise, the null hypothesis is not rejected.

One- and two-tailed critical values are given in Table A.7 in Appendix A. These values vary with the sample size n and level of significance α.

3.1.3   Calculation of Critical Values: Example

Consider a sample of size $n = 3$, and assume that there are no tied x or y observations. Let 1, 2, and 3 be the order of ranks for X variable. There are $3! = 6$ possibilities for the corresponding order of ranks for Y variable. These ranks, their difference, and the calculated Spearman correlation coefficient are presented in the table below. Note that since $n = 3$, the correlation coefficient is computed according to the formula

$$r_s = 1 - \frac{6\sum_{i=1}^{n}(x_i - y_i)^2}{n^3 - n} = 1 - \frac{\sum_{i=1}^{3}(x_i - y_i)^2}{4}.$$

X	Y	$X-Y$	r_s
1	1	0	
2	2	0	$1 - 0 = 1$
3	3	0	
1	1	0	
2	3	-1	$1 - 2/4 = 1/2$
3	2	1	
1	2	-1	
2	1	1	$1 - 2/4 = 1/2$
3	3	0	
1	2	-1	
2	3	-1	$1 - 6/4 = -1/2$
3	1	2	
1	3	-2	
2	1	1	$1 - 6/4 = -1/2$
3	2	1	
1	3	-2	
2	2	0	$1 - 8/4 = -1$
3	1	2	

The upper-tailed probabilities for r_s are $\mathbb{P}(r_s \geq 1) = 1/6$ and $\mathbb{P}(r_s \geq 1/2) = 3/6 = 1/2$. For larger sample sizes, some of the upper-tailed probabilities are smaller than 0.05 or 0.01. The corresponding values are tabulated as one-tailed critical values.

3.1.4   SAS Implementation

There are two commonly used ways to compute the Spearman rank correlation coefficient and the corresponding P-value for testing its equality to zero. SAS outputs the exact P-value in one case and an approximate one in the other.

(i) The FREQ procedure with the EXACT SCORR statement may be used to calculate the coefficient and the corresponding <u>exact</u> P-value. The syntax is:

```
PROC FREQ DATA=data_name;
    TABLE variable_name1*variable_name2;
        EXACT SCORR;
RUN;
```

• The EXACT statement requests the exact p-value.
• The specification SCORR refers to the Spearman rank correlation coefficient.

(ii) The CORR procedure may be applied to compute the coefficient and the corresponding two-sided asymptotic P-value of the test. The syntax is:

```
PROC CORR DATA=data_name SPEARMAN;
    VAR variable_name1 variable_name2;
RUN;
```

• The SPEARMAN option must be specified to request the Spearman correlation. By default, only the Pearson correlation coefficient is calculated.
• The VAR statement lists the variables between which the correlation coefficient will be computed. If this statement is omitted, SAS outputs an array of correlation coefficients for all pairs of variables in the data set.
• The P-value is computed based on the asymptotic approximation, according to which the statistic $T = r_s \sqrt{\dfrac{n-2}{(1-r_s^2)}}$ has an approximate t-distribution with $n-2$ degrees of freedom.

3.1.5   Examples

Example 3.1 In Example 1.1, the IOP reduction in treated and control eyes was measured for nine glaucoma patients. The ophthalmologist predicts that there is a positive correlation between the measurements. To verify the claim, we compute the Spearman rank correlation coefficient and test whether it is significantly larger than zero. The ranked observations and the difference between the ranks are summarized in the following table:

Patient Number	IOP Reduction in Tx Eye	Rank, x	IOP Reduction in Cx Eye	Rank, y	Difference $x - y$
1	0.45	3	0.38	5	-2
2	1.95	9	0.90	8	1
3	1.20	7	0.70	7	0
4	0.65	5	-0.50	2	3
5	0.98	6	0.47	6	0
6	-1.98	1	-1.30	1	0
7	1.80	8	1.34	9	-1
8	-0.76	2	0.13	4	-2
9	0.56	4	-0.40	3	1

There are no ties in the data, therefore, (3.3) with $n = 9$ is used to calculate the correlation coefficient,

$$r_s = 1 - \frac{6\sum_{i=1}^{n}(x_i - y_i)^2}{n^3 - n} = 1 - \frac{(6)(20)}{720} = 0.8333.$$

For $\alpha = 0.01$ and $n = 9$, a one-sided critical value is equal to 0.783. The value of r_s is larger than the critical value, therefore, the null hypothesis is rejected, and the conclusion is that there is a significant positive correlation present.

We arrive at the same conclusion after running these lines of code in SAS:

```
proc freq data=glaucoma;
    table Tx*Cx;
        exact scorr;
run;
```

The output is:

```
Spearman Correlation Coefficient
Correlation (r) 0.8333
Exact Test
One-sided Pr >= r 0.0041
```

Alternatively, the CORR procedure may be run to obtain an approximate P-value. The code is

```
proc corr data=glaucoma spearman;
    var Tx Cx;
run;
```

The results are

```
Spearman Correlation Coefficients, N = 9
        Prob > |r| under H0: Rho=0
                    Tx              Cx
        Tx      1.00000         0.83333
                                0.0053
        Cx      0.83333         1.00000
                0.0053
```

From the outputs, the one-sided exact P-value $= 0.0041$ and one-sided approximate P-value $= 0.0053/2 = 0.0027$ are both smaller than 0.01. This indicates that the null hypothesis is rejected at the 1% significance level, and the conclusion supporting the ophthalmologist's prediction of positive correlation between the measurements is drawn. $\square$

Example 3.2 Researchers in health sciences conjecture that there is a negative association between cigarette smoking and health. They survey ten adults using random-digit dialing method and record the number of years that the respondents smoked cigarettes regularly and their general health condition (poor (1), fair (2), good (3), or excellent (4)). The data and their ranks are:

No.	Years Smoked	Rank, x	Health Condition	Rank, y	Difference, $x - y$
1	12	7.5	1	1.5	6
2	33	9	2	4	5
3	5	5	3	7	-2
4	3	3.5	3	7	-3.5
5	12	7.5	2	4	3.5
6	6	6	2	4	2
7	3	3.5	4	9.5	-6
8	0	1	4	9.5	-8.5
9	2	2	3	7	-5
10	46	10	1	1.5	8.5

Tied observations are present for both X and Y variables. There are two sets of equal observations for X: there are two 3's and two 12's, hence, $T_1 = T_2 = 2$, and $T_x = 2(2^3 - 2) = 12$. For Y, there are two 1's and 4's, and three 2's and 3's, thus, $T_y = 2(2^3 - 2) + 2(3^3 - 3) = 60$. The sum of squares of the differences is $\sum_{i=1}^{10}(x_i - y_i)^2 = 299$. To compute the Spearman correlation coefficient, we use formula (3.4) with $n = 10$. We have

$$r_s = \frac{n^3 - n - 6\sum_{i=1}^{n}(x_i - y_i)^2 - (T_x + T_y)/2}{\sqrt{(n^3 - n)^2 - (T_x + T_y)(n^3 - n) + T_x T_y}}$$

$$= \frac{10^3 - 10 - (6)(299) - (12 + 60)/2}{\sqrt{(10^3 - 10)^2 - (12 + 60)(10^3 - 10) + (12)(60)}} = -0.88078.$$

The alternative hypothesis in this example is lower-tailed, $H_1 : \rho_s < 0$, since negative association is conjectured. The critical value from Table A.7 for $n = 10$ and one-tailed $\alpha = 0.01$ is 0.745. The absolute value of the test statistic $|r_s| = 0.88078$ is above the critical value, therefore, the null hypothesis is rejected at the 1% significance level, and we say that negative association between smoking and health exists.

In SAS, to run this test with the exact P-value, we enter the code:

```
data smoking_n_health;
   input yrs_smoked health @@;
datalines;
12 1 33 2 5 3 3 3 12 2 6 2 3 4 0 4 2 3 46 1
;
proc freq data=smoking_n_health;
   table yrs_smoked*health;
      exact scorr;
run;
```

The relevant output is

```
Spearman Correlation Coefficient
Correlation (r) -0.8808
Exact Test
One-sided Pr <= r 9.524E-04
```

To run this test with an asymptotic P-value, we use the following code:

```
proc corr data=smoking_n_health spearman;
   var yrs_smoked health;
run;
```

The output is

```
Spearman Correlation Coefficients, N = 10
        Prob > |r| under H0: Rho=0
                        yrs_
                        smoked          health
   yrs_smoked           1.00000        -0.88078
                                        0.0008
   health              -0.88078         1.00000
                        0.0008
```

Here, the one-sided exact P-value $= 0.0009524$, whereas the one-sided approximate P-value $= 0.0008/2 = 0.0004$. Since both are much smaller than 0.01, the null hypothesis is rejected at the 1% significance level. The conclusion is the same as above. □

Example 3.3 An instructor of a graduate course in applied statistics suspects that there is no correlation between how many hours students study for an exam and their scores on that exam. She surveys her 14 students and records how many hours they spent studying for a midterm exam. She also records their scores on the exam. The data along with ranks are given in the following table:

No.	Hours Studied	Rank, x	Exam Score	Rank, y	Difference, $x-y$
1	0	1	28	1	0
2	5	3.5	94	12	-8.5
3	9	8	84	9	-1
4	7	6	45	2	4
5	17	11.5	82	8	3.5
6	17	11.5	99	14	-2.5
7	5	3.5	67	5	-1.5
8	12	10	97	13	-3
9	21	13	79	7	6
10	3	2	93	11	-9
11	7	6	62	4	2
12	29	14	60	3	11
13	7	6	85	10	-4
14	10	9	78	6	3

Tied observations are present in the x data, but not y data. There are two 5's, two 17's, and three 7's for the X variable. Thus, $T_1 = T_2 = 2$, $T_3 = 3$, and $T_x = 2(2^3 - 2) + (3^3 - 3) = 36$. Also, $n = 14$, $\sum_{i=1}^{14}(x_i - y_i)^2 = 386$, and $T_y = 0$. The expression in (3.4) is used to compute the correlation coefficient,

$$r_s = \frac{n^3 - n - 6\sum_{i=1}^{n}(x_i - y_i)^2 - (T_x + T_y)/2}{\sqrt{(n^3 - n)^2 - (T_x + T_y)(n^3 - n) + T_x T_y}}$$

$$= \frac{14^3 - 14 - (6)(386) - (36+0)/2}{\sqrt{(14^3 - 14)^2 - (36+0)(14^3 - 14) + (36)(0)}} = 0.14602.$$

The alternative hypothesis is two-sided, $H_1 : \rho_s \neq 0$. From Table A.7, the two-tailed critical value that corresponds to $\alpha = 0.05$ and $n = 14$ is 0.538. The correlation coefficient is smaller than the critical value, therefore, we fail to reject the null hypothesis and conclude that the number of hours spent studying for an exam and exam scores are not correlated.

In SAS, to obtain the exact P-value, we run the code:

```
data studying4exam;
   input hours score @@;
datalines;
0 28   5 94 9 84 7 45 17 82 17 99 5 67
12 97 21 79 3 93 7 62 29 60 7 85 10 78
;

proc freq data=studying4exam;
   table hours*score;
      exact scorr;
```

```
run;
```

That produces the output:

```
Spearman Correlation Coefficient
Correlation (r) 0.1460
Exact Test
Two-sided Pr >= |r| 0.6162
```

For an approximate P-value, we run the CORR procedure

```
proc corr data=studying4exam spearman;
    var hours score;
run;
```

The output is

```
Spearman Correlation Coefficients, N = 14
        Prob > |r| under H0: Rho=0
                    hours            score
    hours         1.00000          0.14602
                                    0.6184
    score         0.14602          1.00000
                  0.6184
```

Thus, the two-sided exact P-value is 0.6162, and an approximate one is 0.6184. Both probabilities are much larger than 0.05, which supports our conclusion.   □

3.2   Fisher Exact Test

Consider two categorical variables one of which is necessarily nominal and the other is either nominal or ordinal. There are two situations when measurements of these variables may be written in the form of a *contingency table* (also called *cross tabulation* or *crosstab* or *two-way table*) with r rows and c columns. The first scenario is when r samples are drawn from independent populations and a variable with c levels is observed for each individual. The frequencies for every sample and every level of the variable may be summarized in an $r \times c$ table. The second case is when a single sample is drawn from one population and two variables with r and c levels, respectively, are observed for each individual. The data may be presented in the form of an $r \times c$ table of frequencies for each level-combination of these two variables.

If observed frequencies are sufficiently high, a chi-square test can be carried out in both situations. The testing procedures are identical but statistical hypotheses differ. For several samples, equality of proportions is tested, that is, the null hypothesis is that the column proportions are equal across the rows (which represent samples), and the alternative hypothesis is that column proportions are not all equal. For a single sample, the null hypothesis is that there is no association between the measured variables (i.e., they are independent), whereas the alternative hypothesis states that an association exists.

The assumption of validity of the chi-square test is that the expected count under H_0 for each cell in a contingency table is at least 5. The expected count in a cell is defined as the product of the corresponding marginal totals divided by the grand total. If this assumption is violated and some expected counts are below 5, a nonparametric *Fisher exact test* may be used as a replacement. The test bears the name of Sir Ronald Aylmer Fisher (1890-1962), a famous English statistician. He was the first who introduced this test for a 2×2 contingency table.[1] It was later extended to a general $r \times c$ table.[2]

3.2.1 Testing Procedure

Calculations for this test are too cumbersome to do by hand, and are typically carried out by a software. Firstly, the marginal totals are computed for each of the r rows and c columns. Secondly, all possible tables with the same marginal totals are enumerated, and the probability of observing each table is computed. Finally, the P-value for the Fisher exact test is calculated as the sum of all probabilities that are less than or equal to the probability of the actually observed table. If the P-value is larger than a pre-specified level α, it indicates that the observed table is likely under the equality of proportions (or independence, if applicable) assumption, and the null hypothesis cannot be rejected. If, on the other hand, the P-value is less than or equal to α, this means that the observed table is unusual under H_0 and the null should be rejected.

Next, we give an explanation of how to compute the probability of a table with fixed marginal totals in the case of a 2×2 table. For higher dimensions of a contingency table, analogous formulas may be derived.

Consider a 2×2 contingency table, and assume that the row and column totals and the overall total are pre-determined.

	Column 1	Column 2	Total
Row 1	a	b	$a+b$
Row 2	c	d	$c+d$
Total	$a+c$	$b+d$	$a+b+c+d$

The number of such tables with the cell frequencies in the first row equal to a and b, given the fixed column sums, is $\binom{a+c}{a}\binom{b+d}{b}$. The overall number of tables with

---

[1]Fisher, R. A. (1925) *Statistical Methods for Research Workers*, Oliver and Boyd.
[2]Freeman, G.H. and Halton, J.H. (1951) Note on an exact treatment of contingency, goodness of fit and other problems of significance, *Biometrika*, **38**, 141-149.

the first row sum of $a+b$ is $\binom{a+b+c+d}{a+b}$. Therefore, the probability to observe this table is

$$p = \frac{\binom{a+c}{a}\binom{b+d}{b}}{\binom{a+b+c+d}{a+b}} = \frac{(a+b)!(c+d)!(a+c)!(b+d)!}{(a+b+c+d)!a!b!c!d!}. \tag{3.5}$$

3.2.2   Calculation of P-values: Example

As an example, suppose the following 2×2 contingency table is observed:

	Column 1	Column 2	Total
Row 1	4	1	5
Row 2	2	9	11
Total	6	10	16

According to (3.5), the probability of observing this table is

$$\frac{5!\,11!\,6!\,10!}{16!\,4!\,1!\,2!\,9!} = 0.0343.$$

There are a total of six tables with the same marginal sums. They are listed below along with the associated probabilities. The tables that have probabilities of 0.0343 or smaller are marked with an asterisk. The P-value is the sum of these probabilities.

Table		Probability	
0	5	$\dfrac{5!\,11!\,6!\,10!}{16!\,0!\,5!\,6!\,5!}$	$= 0.0577$
6	5		
1	4	$\dfrac{5!\,11!\,6!\,10!}{16!\,1!\,4!\,5!\,6!}$	$= 0.2885$
5	6		
2	3	$\dfrac{5!\,11!\,6!\,10!}{16!\,2!\,3!\,4!\,7!}$	$= 0.4121$
4	7		
3	2	$\dfrac{5!\,11!\,6!\,10!}{16!\,3!\,2!\,3!\,8!}$	$= 0.2060$
3	8		
4	1	$\dfrac{5!\,11!\,6!\,10!}{16!\,4!\,1!\,2!\,9!}$	$= 0.0343^*$
2	9		
5	0	$\dfrac{5!\,11!\,6!\,10!}{16!\,5!\,0!\,1!\,10!}$	$= 0.0014^*$
1	10		

The P-value for the Fisher exact test in this example is $0.0343 + 0.0014 = 0.0357$.

3.2.3   SAS Implementation

Fisher exact test may be requested in SAS as an option in the PROC FREQ statement that presents observed frequencies for nominal or ordinal scale variables in a crosstab.

• For a data set containing one row for each individual, the syntax is:

```
PROC FREQ DATA=data_name;
    TABLE  variable_name1*variable_name2 / FISHER;
RUN;
```

Alternatively, the EXACT statement may be used:

```
PROC FREQ DATA=data_name;
   TABLE  variable_name1*variable_name2;
      EXACT FISHER;
RUN;
```

• For a data set containing level combinations and cell counts, the syntax is:

```
PROC FREQ DATA=data_name ORDER=DATA;
   WEIGHT count;
      TABLE  variable_name1*variable_name2 / FISHER;
RUN;
```

Below the contingency table, under the title Fisher's Exact Test, SAS outputs the P-value marked Prob <= P. In the case of a 2×2 table, the P-value is called Two-sided Pr <= P.

3.2.4   Examples

Example 3.4 An administration of a college is interested in increase of utilization of a student gym. They randomly sample 15 students and record their level of daily physical activity (high/moderate/low) and whether they exercise in the gym (yes/no). The data are

Student	Activity Level	Gym Used
1	high	yes
2	moderate	yes
3	low	yes
4	low	no
5	low	no
6	low	yes
7	low	no
8	moderate	yes
9	moderate	yes
10	high	no
11	low	no
12	moderate	yes
13	moderate	yes
14	high	yes
15	moderate	yes

To test H_0 that there is no association between these two variables, we conduct the Fisher exact test by running the following SAS code:

```
data student_fitness;
      input actlevel $ gym_used $ @@;
   datalines;
high no moderate no low yes low yes low yes low yes
low no moderate yes moderate yes high no low no
moderate yes moderate yes high yes moderate yes
;

proc freq data=student_fitness;
   table actlevel * gym_used/fisher;
run;
```

As part of the output, SAS computes the P-value for the Fisher exact test.

```
 Fisher's Exact Test
Pr <= P        0.3357
```

The P-value is larger than 0.05, therefore, we fail to reject the null hypothesis and conclude that there is no association between students' fitness level and gym utilization. □

Example 3.5 A study is conducted on patients with psoriasis, an inflammatory skin disease. Two new drugs (A and B) are tested for efficacy of relieving psoriasis itch. Thirty patients are eligible for the study. They are randomly evenly assigned to one of the two treatment groups or the placebo group. Three months later, two patients in the placebo group, eight patients in the drug A group, and six patients in the drug B group have experienced a relief. Since cell counts are small, we employ the Fisher exact test to see whether proportions of patients with relief are the same in all three groups. To carry out the test in SAS, first a data set must be created. There are two alternative ways to do so. The first one is by using the DO loops. The code is given below.

```
data psoriasis1;
    do counter=1 to 2;
        relif='yes';
          group='Placebo';
        output;
    end;
        do counter=1 to 8;
            relief='no';
              group='Placebo';
        output;
        end;

    do counter=1 to 8;
        relief='yes';
          group='Drug A';
        output;
    end;
        do counter=1 to 2;
            relief='no';
              group='Drug A';
        output;
        end;

    do counter=1 to 6;
        relief='yes';
          group='Drug B';
        output;
    end;
        do counter=1 to 4;
            relief='no';
              group='Drug B';
        output;
        end;
run;

proc freq data=psoriasis1;
    table relief*group/fisher;
```

```
run;
```

The second way to create a data set is by specifying cell counts. This is how it is done:

```
data psoriasis2;
   input   group $ 1-7 relief $ count;
cards;
Placebo yes 2
Placebo no  8
Drug A  yes 8
Drug A  no  2
Drug B  yes 6
Drug B  no  4
;

proc freq data=psoriasis2 order=data;
    weight count;
      tables relief*group/fisher;
run;
```

The P-value in this example is

```
Fisher's Exact Test
Pr <= P 0.0366
```

Based on SAS output, the P-value is 0.0366. It is between 0.01 and 0.05, thus, we can reject the null hypothesis at the 5% significance level and conclude that proportions of patients who experienced relief differ across the groups. At the 1%, however, we conclude that the proportions are the same.   □

Exercises for Chapter 3

Exercise 3.1 A student in marketing wants to see for himself which products increase in price as gasoline prices increase. He finds online average annual prices for a number of products for a certain geographical area he is interested in. In particular, he looks at prices of gasoline vs. prices of milk for the past 12 years. The average prices per gallon, adjusted for inflation, are:

Price of Gas Per Gallon	Price of Milk Per Gallon
$1.78	$1.30
$2.11	$1.70
$2.01	$1.88
$2.17	$2.15
$2.45	$2.20
$2.76	$2.25
$3.12	$2.19
$3.24	$2.45
$3.56	$2.87
$3.70	$2.99
$3.42	$3.15
$3.24	$3.06

Compute the Spearman rank correlation coefficient. Test whether milk prices and gasoline prices increase together. Clearly state your hypotheses and conclusion. Do calculations by hand and in SAS. In SAS compute both exact and approximate P-values.

Exercise 3.2 A family therapist conjectures that there is a negative association between the highest educational degree a husband attains and the severity of domestic violence. She surveys her next 15 clients and collects information on both variables. Education variable is broken into three categories: less than high school (<HS), high school graduate or equivalent (HS grad), and some college or higher (HS+). Severity of violence is measured at three levels: never, sometimes, and often. The data are given below.

Education	Violence
HS grad	never
HS+	often
HS grad	often
HS grad	sometimes
HS+	never
HS+	never
HS+	often
HS grad	often
<HS	often
HS+	never
<HS	often
<HS	never
HS grad	sometimes
HS+	sometimes
<HS	sometimes

Is there enough evidence in these data to support the conjecture? Compute the Spearman correlation coefficient and conduct the appropriate test by hand and in SAS. In SAS calculate an exact P-value as well as an asymptotic one.

Exercise 3.3 A basketball fan records the number of points that Kobe Bryant scored during some 11 years. The data are:

Year	Points
2000	1938
2001	2019
2002	2461
2003	1557
2004	1819
2005	2832
2006	2430
2007	2323
2008	2201
2009	1970
2010	2078

Compute the Spearman correlation coefficient and carry out the test to see whether these two variables are correlated. State your hypotheses and conclusion. Repeat the analysis in SAS. Calculate both exact and t-approximate P-values.

Exercise 3.4 A pilot study of a new heart valve with a total of 90 patients is conducted at three centers: in Stockholm, Washington, and Denver. Investigators suspect that patient populations in these three centers differ with respect to age. The data are as follows:

Center	Patient Age at Surgery			
	20s and 30s	40s and 50s	60s	70s and above
Stockholm	5	10	13	12
Washington	8	11	4	3
Denver	2	14	8	0

Confirm or dissipate the investigators' suspicion by conducting the Fisher exact test. Clearly state the null and the alternative hypotheses and your conclusion.

Exercise 3.5 A school psychologist picks at random 50 middle schoolers and asks them whether studying, sports, or just hanging out with friends is more important to them. He records their answers and presents his findings classified by gender in a two-way table:

Gender	Studying	Sports	Friends
Boy	12	11	4
Girl	8	2	13

Formulate the hypotheses that the psychologist might be interested in checking and carry out the Fisher exact test.

Exercise 3.6 A library science specialist is wondering whether American fiction writers were more likely to use pen names before the 20th century. She randomly selects 30 fiction writers, 15 who wrote before the 20th century and 15, after the 20th century, and finds that six of the writers in the first group used pen names, and nine used their real names, whereas in the second group, four wrote under a pen name and eleven used real names. Formulate statistical hypotheses that reflect the specialist's interest. Conduct the Fisher exact test. When applying SAS,

(a) use the data set created with the DO loop and containing one row for each writer.

(b) use the data set containing level combinations and cell counts.

Exercise 3.7 A large anthropological survey was conducted among three indigenous tribes of the Amazon rainforest. Among other questions, investigators asked to identify the respondent's major occupation. The responses were classified into five groups: farming, hunting, gathering, fishing, and trading. The data given below are a subsample of a larger data set.

Tribe	Occupation	Tribe	Occupation	Tribe	Occupation
A	farming	B	farming	C	farming
A	hunting	B	gathering	C	gathering
A	farming	B	gathering	C	gathering
A	gathering	B	fishing	C	trading
A	gathering	B	farming	C	farming
A	farming	B	fishing	C	fishing
A	trading	B	trading	C	farming
A	farming	B	hunting	C	gathering
A	farming	B	farming	C	fishing
A	farming	B	farming	C	fishing
A	farming	B	farming	C	gathering
A	gathering	B	farming	C	fishing
A	fishing	B	gathering	C	farming
A	farming	B	fishing	C	farming
A	fishing	B	trading	C	farming
A	farming	B	trading	C	fishing
A	hunting	B	fishing	C	gathering
A	gathering	B	fishing	C	trading
A	hunting	B	hunting	C	trading
A	trading	B	fishing	C	farming

Write down the hypotheses and carry out the Fisher exact test. Present the contingency table and P-value. State your conclusion clearly.

Chapter 4

Nonparametric Regression

For some data sets it is impossible to describe the relation between the response variable y and predictor variables $x_1, \ldots, x_k$ by a function of an a priori known form (for example, polynomial or exponential). A useful tool in this case is a *nonparametric* regression, specified by

$$y = f(x_1, \ldots, x_k) + \varepsilon \tag{4.1}$$

where f is a nonparametric *response function* with no explicit functional form, and ε's are independent identically distributed random errors with a zero mean and constant variance σ^2. No assumption is made on the form of the probability distribution of ε's.

As an illustration, consider the following scatterplot of a response y against a predictor x. The pattern on the graph shows periodic spikes (sometimes termed *cycles*, *periodic variations* or *seasonal variations*). None of the traditional response functions from the parametric world would have a satisfactory fit to these data.

A scatterplot of response Y against predictor X

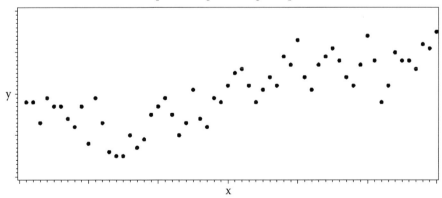

Nonparametric regression methods estimate the response function f and fit a curve (or surface in three or higher dimensions) on the scatterplot. In this chapter we discuss two methods of estimation: the loess regression and thin-plate splines.

4.1   Loess Regression

4.1.1   Definition

Suppose n sets of observations $(x_1,\dots,x_k,y)$ are available where the relation (4.1) holds. The loess (*locally estimated scatterplot smoothing*) method evaluates f at every data point and plots on the scatterplot a piece-wise polynomial curve connecting the fitted points. In more than two dimensions, the loess regression approach constructs flat surfaces through the fitted points. This method was introduced in 1979 by William S. Cleveland (1943-), an American statistician.[1]

The estimation of the response function at the data points is done by fitting a polynomial regression function in the local neighborhood of each point. The radius of the neighborhoods is determined by a pre-specified fraction of data points, called *smoothing parameter*, that these neighborhoods encompass. A polynomial regression (typically linear or quadratic) is fit by the weighted least-squares method with less weight given to points more distant from the center.

Mathematically speaking, this is how the fitting in local neighborhoods is done. First, the value of the smoothing parameter is specified as, say, p/n. It means that for each fixed point $\mathbf{x}^0 = (x_1^0,\dots,x_k^0)$, a linear function is fit through the p points in the local neighborhood $\mathcal{N}_p(\mathbf{x}^0)$. This linear function is defined as

$$l(\mathbf{x}) = l(x_1,\dots,x_k) = \beta_0 + \beta_1(x_1 - x_1^0) + \cdots + \beta_k(x_k - x_k^0),$$

for every $\mathbf{x} = (x_1,\dots,x_k)$ in $\mathcal{N}_p(\mathbf{x}^0)$.

The weighted least-squares predicted line $\hat{l}(\mathbf{x}) = \hat{\beta}_0 + \hat{\beta}_1(x_1 - x_1^0) + \cdots + \hat{\beta}_k(x_k - x_k^0)$ minimizes

$$\sum_{\mathbf{x}_i \in \mathcal{N}_p(\mathbf{x}^0)} \left[y_i - l(\mathbf{x}_i) \right]^2 w\left(\frac{\|\mathbf{x}_i - \mathbf{x}^0\|}{r(\mathbf{x}^0)} \right)$$

where $\mathbf{x}_i = (x_{1i},\dots,x_{ki}), i = 1,\dots,p$. Here $\|\mathbf{x}_i - \mathbf{x}^0\|$ denotes the Euclidean distance, that is,

$$\|\mathbf{x}_i - \mathbf{x}^0\| = \sqrt{(x_{1i} - x_1^0)^2 + \cdots + (x_{ki} - x_k^0)^2},$$

$r(\mathbf{x}^0) = \max\limits_{\mathbf{x}_i \in \mathcal{N}_p(\mathbf{x}^0)} \|\mathbf{x}_i - \mathbf{x}^0\|$ is the radius of the neighborhood, and $w(\cdot)$ is the weight function.

Finally, the response function f is estimated at the point $\mathbf{x}^0$ by the corresponding value on the predicted line $\hat{l}$, that is, $\hat{f}(\mathbf{x}^0) = \hat{l}(\mathbf{x}^0) = \hat{\beta}_0$. Note that only the value of the fitted intercept comes into play.

---

[1] Cleveland, W. S. (1979) Robust locally weighted regression and smoothing scatterplots, *Journal of the American Statistical Association*, **74**, 829-836.

Once the loess estimation of the response function at every data point is completed, the loess curve is drawn on a scatterplot by connecting the fitted points with straight lines.

If a considerable curvilinear relation is displayed on a scatterplot, a quadratic function

$$q(x_1,\ldots,x_k) = \beta_0 + \beta_{1,1}(x_1 - x_1^0) + \beta_{1,2}(x_1 - x_1^0)^2 + \cdots + \beta_{k,1}(x_k - x_k^0) + \beta_{k,2}(x_k - x_k^0)^2$$

is fitted instead of the linear function l.

Different weight functions have been proposed for the loess regression method. The most commonly used ones are

- the *bisquare* function

$$w(\mathbf{x}) = \begin{cases} \left(1 - \|\mathbf{x}\|^2\right)^2, & \text{if } \|\mathbf{x}\| \leq 1, \\ 0, & \text{otherwise,} \end{cases}$$

and

- the *tricube* function

$$w(\mathbf{x}) = \begin{cases} \left(1 - \|\mathbf{x}\|^3\right)^3, & \text{if } \|\mathbf{x}\| \leq 1, \\ 0, & \text{otherwise.} \end{cases}$$

In both cases $\|\mathbf{x}\|$ denotes the Euclidean norm of $\mathbf{x} = (x_1,\ldots,x_k)$, $\|\mathbf{x}\| = \sqrt{x_1^2 + \cdots + x_k^2}$.

Going back to our example, the figure below displays a series of scatterplots with the fitted loess curve for different values of the smoothing parameter. The curve with the smallest value of the smoothing parameter simply zigzags connecting the data points. As the value of the smoothing parameter increases, the fitted curve becomes smoother. First it nicely captures the periodic variations but then becomes more and more stretched out.

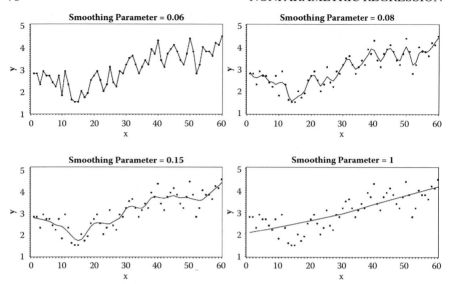

This behavior is not surprising. If the value of the smoothing parameter is close to zero, then, when fitting a polynomial regression, the local neighborhood is assigned only one point, and that is its center. So the fitted value of f coincides with the observed value, and the loess curve just connects the observed points. At the other extreme, when the smoothing parameter equals the unity, all observed points are included in all neighborhoods, and, therefore, a straight line is always fit through all points. The fit, however, is weighted, so it may not always result in a perfectly straight-line loess curve. If the value of the smoothing parameter is chosen larger than one, the loess curve eventually becomes a straight line. In some cases, depending on configuration of points on a scatterplot, the loess curve may already look perfectly smooth even for relatively small values of the smoothing parameter.

In this figure, the loess curves are plotted for smoothing parameters 0.06, 0.08, 0.15, and 1. Notice how the behavior of the curves ranges from sequentially connecting all points to becoming absolutely flat. To capture properly the periodicity in the data, using a value of the smoothing parameter somewhere in-between 0.08 and 0.15 may be recommended.

4.1.2 Smoothing Parameter Selection Criterion

Many criteria have been proposed for selection of an optimal smoothing parameter for a set of data. Here we explain the most frequently used, the *bias corrected Akaike information criterion* (AICC). An information criterion was originally introduced by a Japanese statistician Hirotugu Akaike (1927-2009) in 1974.[1] It was extended to the

---

[1] Akaike, H. (1974) A new look at the statistical model identification, *IEEE Transactions on Automatic Control*, **19**, 716-723.

bias corrected version in 1989.[2] The criterion is to select the smoothing parameter that minimizes

$$AICC = \ln \hat{\sigma}^2 + \frac{2(Trace(\mathbb{L}) + 1)}{n - Trace(\mathbb{L}) - 2}$$

where $\hat{\sigma}^2$ is an estimate of the variance of random error ε in (4.1). The matrix $\mathbb{L}$ is an $n \times n$ *smoothing matrix* that satisfies $\mathbf{y} = \mathbb{L}\hat{\mathbf{y}}$ where $\mathbf{y}$ is the vector of observed *y*-values, and $\hat{\mathbf{y}}$ is the vector of the corresponding predicted values. The trace of a matrix is defined as the sum of the elements on the main diagonal.

4.1.3 SAS Implementation: Fitting Loess Regression

In SAS, the LOESS procedure is used to fit nonparametric curves (or surfaces in higher dimensions). The basic syntax is:

```
PROC LOESS DATA=data_name;
    MODEL response=list of predictors/
       DEGREE=1 or 2 CLM ALPHA=value SMOOTH=values;
   ODS OUTPUT OutputStatistics=output_stats_data_name
       ScoreResults=score_results_data_name;
          SCORE DATA=points4prediction_data_name;
RUN;
```

• The option DEGREE= specifies the degree of local polynomials (1 for linear, 2 for quadratic). The default value is 1.
• The CLM option (stands for "confidence limits for the mean") produces a confidence interval for every observed response value. By default, 95% limits are computed, but the level can be changed by using the ALPHA= option.
• As a default, SAS uses the bias corrected Akaike Information Criterion (AICC) to select an optimal smoothing parameter. If the SMOOTH= option is included in the code, a separate loess regression is fitted for every listed value of the smoothing parameter.
• The ODS OUTPUT OutputStatistics statement creates an output data set containing the results of fitting a loess regression. Here ODS stands for "output delivery system." The output data set contains a column of the values of the smoothing parameter used in the estimation called SmoothingParameter, a column of the response variable values called DepVar ("dependent variable"), a column of the fitted values called Pred ("predicted"), a column of the lower confidence limits LowerCL, and a column of the upper confidence limits UpperCL.
• The SCORE statement specifies a data set with points for which obtaining predicted values is desired. These values are placed in the data set given in the ScoreResults statement, and can be viewed by printing this file. The last column in the data containing scoring results, called *P_response*, is the column of predicted values.

---

[2]Hurvich, C.M. and Tsai, C.-L. (1989) Regression and time series model selection in small samples, *Biometrika*, **76**, 297-307.

• For local fitting, SAS uses the normalized tricube weight function defined as

$$w(\mathbf{x}) = \begin{cases} \frac{32}{5}\left(1 - \|\mathbf{x}\|^3\right)^3, & \text{if } \|\mathbf{x}\| \le 1, \\ 0, & \text{otherwise.} \end{cases}$$

4.1.4   SAS Implementation: Plotting Fitted Loess Curve

To plot a fitted loess curve on a two-dimensional scatterplot, apply the following syntax:

```
SYMBOL1 COLOR=black VALUE=dot; /*black dots on scatterplot*/
SYMBOL2 COLOR=black VALUE=none INTERPOL=join LINE=1;
/*curve displayed as solid */
SYMBOL3 COLOR=black VALUE=none INTERPOL=join LINE=2;
/*LowerCL band is displayed as dashed*/
SYMBOL4 COLOR=black VALUE=none INTERPOL=join LINE=2;
/*UpperCL band is displayed as dashed*/

PROC GPLOT data=output_data_name;
    BY SmoothingParameter;
        PLOT (DepVar Pred LowerCL UpperCL)*predictor / OVERLAY
            NAME='graph_name';
RUN;

PROC GREPLAY IGOUT=WORK.GSEG TC=SASHELP.TEMPLT
        TEMPLATE=template_name NOFS;
            TREPLAY 1:graph_name 2:graph_name1 ... ;
RUN;
```

• The BY statement is required for producing a scatterplot with the fitted loess curve for every value of the smoothing parameter specified earlier in the code. The value of the corresponding smoothing parameter is displayed on each graph.

• The option OVERLAY in the PLOT statement requests all the listed entities (DepVar, Pred, LowerCL, and UpperCL) to be displayed on the same graph.

• The option NAME in the PLOT statement assigns a name to each graph. The first graph is called *graph_name*. All consecutive graphs have the same name but added suffixes 1,2,3, etc.

• The procedure GREPLAY (stands for "graphics replay") displays the graphs in a certain layout provided by a template. The option IGOUT tells SAS that the graphs are located in the WORK directory in the default graph catalog GSEG. The TC ("template catalog") directs SAS to the folder SASHELP.TEMPLT where templates are located. The option TEMPLATE selects a particular template. The option NOFS (" no full-screen") allows the graphs to be displayed properly. The TREPLAY ("template replay") statement positions the graphs according to the chosen template. The most frequently used templates are 12r2 (two by two array: two boxes on the left and two on the right) and 12r2s (same as 12r2s but with a space in-between the boxes).

4.1.5   SAS Implementation: Plotting 3D Scatterplot

The syntax for plotting a 3D scatterplot is

```
PROC G3D data=data_name;
    SCATTER predictor1*predictor2=response/ COLOR="color";
RUN;
```

• The option COLOR specifies the color of the 3D scatterplot. The name of the color must be taken into double quotation marks.

4.1.6   SAS Implementation: Plotting Fitted Loess Surface

SAS will not display a fitted loess surface unless there is enough 3D grid points to draw the planes through. As a remedy, a data set with grid points should be created and included in the SCORE option of the LOESS procedure. The following syntax may be used:

```
DATA grid_points_data_name;
    DO predictor1 = min TO max BY increment;
        DO predictor2 = min TO max BY increment;
          OUTPUT;
        END;
    END;
RUN;
```

After running the LOESS procedure (see Subsection 4.1.3), the P_response variable may be plotted on a 3D graph against two predictors. The syntax is

```
PROC G3D DATA=score_results_data_name;
    PLOT predictor1*predictor2 =P_response/ CTOP=color CBOTTOM=color;
RUN;
```

• The options CTOP and CBOTTOM specify colors for the top and, if visible, bottom parts of the fitted surface.

4.1.7   Examples

Example 4.1 Statistics on monthly unemployment rates for a county has been collected over one decade. The following code creates a scatterplot of the data and fits a loess curve with the optimal values of the smoothing parameter based on the bias corrected Akaike Information Criterion.

```
data unempl_rates;
   input rate @@;
      month=_N_;
datalines;
6.0 6.7 4.9 4.4 5.8 4.8 5.5 6.7 4.7 5.6 6.5 6.0
4.7 5.1 7.2 6.1 7.7 5.7 7.1 4.2 5.8 5.1 6.3 5.1
3.9 4.7 4.4 5.9 4.1 5.8 4.9 5.4 3.9 6.0 4.1 4.6
5.7 5.0 4.5 6.9 5.6 4.6 4.4 4.1 3.2 6.3 4.2 4.7
4.3 4.3 4.5 6.7 3.9 4.6 5.8 3.8 5.5 4.7 5.0 4.2
5.0 4.5 3.7 5.5 5.4 2.6 5.0 4.9 5.7 4.3 5.3 7.1
7.5 4.1 5.1 5.7 4.8 6.1 6.3 4.1 5.7 7.2 6.0 7.2
8.0 8.7 8.5 9.1 7.5 10.5 8.5 7.4 10.5 8.9 8.5 9.9
8.3 9.9 7.2 9.5 10.5 11.9 11.4 8.0 10.5 11.2 9.2 9.5
10.0 10.3 9.1 8.1 7.9 9.5 10.7 8.5 9.1 8.7 9.0 8.6
;

proc loess data=unempl_rates;
   model rate=month/clm;
      ods output OutputStatistics=results;
run;

proc print data=results;
run;

symbol1 color=black value=dot;
symbol2 color=black value=none interpol=join line=1;
symbol3 color=black value=none interpol=join line=2;
symbol4 color=black value=none interpol=join line=2;

proc gplot data=results;
   by SmoothingParameter;
      plot (DepVar Pred LowerCL UpperCL)*month/ overlay;
run;
```

The partial printout of the file results is given below. It contains the values of the predicted response and the 95% lower and upper confidence limits.

Obs	Smoothing Parameter	month	Dep Var	Pred	LowerCL	UpperCL
1	0.2125	1	6.0	5.43046	4.53901	6.32190
2	0.2125	2	6.7	5.46287	4.64449	6.28124
3	0.2125	3	4.9	5.49528	4.74651	6.24405

4	0.2125	4	4.4	5.52769	4.84401	6.21137
5	0.2125	5	5.8	5.56136	4.93570	6.18702
---	lines omitted ---					
116	0.2125	116	8.5	9.00570	8.38004	9.63136
117	0.2125	117	9.1	8.91656	8.23289	9.60024
118	0.2125	118	8.7	8.83072	8.08195	9.57948
119	0.2125	119	9.0	8.74487	7.92649	9.56324
120	0.2125	120	8.6	8.65902	7.76758	9.55046

Here is the scatterplot with the fitted loess curve and the 95% confidence band. The value of the optimal smoothing parameter is 0.2125.

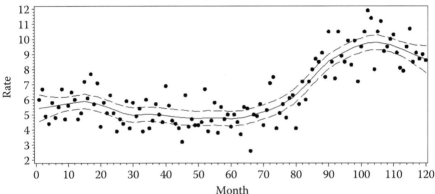

Note that the automatically fitted loess curve doesn't really recognize the periodic trends in the data. We can request from SAS to fit loess curves for certain values of the smoothing parameter by listing those values in the SMOOTH option of the MODEL statement. We run the models and plot the results by typing the following code:

```
proc loess data=unempl_rates;
    model rate=month/ clm smooth=0.05 0.08 0.1 0.12;
    ods output OutputStatistics=results;
run;

proc gplot data=results;
  by SmoothingParameter;
    plot (DepVar Pred LowerCL UpperCL)*month/ overlay name='graph';
  run;

proc greplay igout=work.gseg tc=sashelp.templt template=l2r2 nofs;
    treplay 1:graph 2:graph2 3:graph1 4:graph3;
run;
```

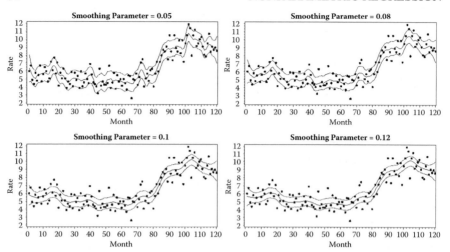

Studying these graphs, we conclude that the smoothing parameter value close to 0.1 describes the periodicity in the data the best. The model, however, won't have the best fit according to the AICC criterion. □

Example 4.2 An urban development corporation considers building a shopping plaza in a town. Prior to commencing construction, the company investigates the buying power and behavior of potential customers. The company's analysts mail out 500 questionnaires, and obtain 58 valid complete responses. They choose three variables to be analyzed: annual family income, annual grocery expenditures, and annual entertainment expenditures. The analysts are interested in defining the population of consumers who spend large amounts of money on entertainment. The data (in thousands of dollars) are given in the following SAS data step:

```
data plaza;
    input income grocery entmt@@;
datalines;
114 15 7 79 23 3 69 8 10 44 8 3.5 18 5 0.8 32 7 2.4
62 18 8 77 10 3 35 10 3.2 66 10 16 71 12 12 25 5 0.6
92 21 5 44 12 4.5 85 15 4 71 18 13 43 13 1.3 78 18 12
89 13 2 116 12 11 82 13 2 41 12 1 37 6 2.4 83 13 1
145 25 12 87 11 12 117 16 8 86 12 2 85 14 3 96 15 3
51 7 1.6 59 13 4 58 10 3.9 46 7 2.4 38 6 0.3 120 22 5
87 10 2 114 17 5 74 11 15 76 10 8 43 8 1.2 37 5 1.2
56 11 1.5 38 8 5.6 28 5 1.2 75 11 16 49 12 1.8 89 10 8
75 20 12 123 22 7 122 25 2 111 12 6 76 10 17 55 7 1
39 6.5 4 46 12 2.8 45 8 2 55 13 9
;
```

The analysts plot entertainment expenditures against the other two variables. The three-dimensional scatterplot is produced by these lines of code:

```
proc g3d data=plaza;
    scatter income*grocery=entmt/ color="black";
run;
```

Here is the graph.

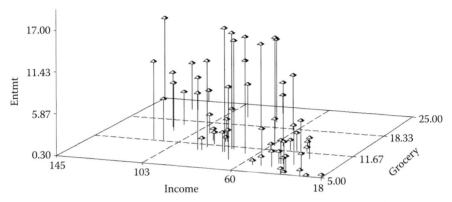

To fit and plot the loess surface, they create a data set containing the grid points that span the range of the data: 18 to 45 for income and 5 to 25 for grocery expenditure.

```
data grid_points;
    do income=18 to 145 by 1;
        do grocery=5 to 25 by 1;
            output;
        end;
    end;
run;
```

Next, the analysts run the loess procedure that creates a file named score_results containing predicted values for all the specified grid points.

```
proc loess data=plaza;
    model entmt=income grocery;
        ods output ScoreResults=score_results;
            score data=grid_points;
run;
```

Lastly, they plot the fitted surface.

```
proc g3d data=score_results;
    plot income*grocery=p_entmt/ctop=black cbottom=gray;
```

```
run;
```

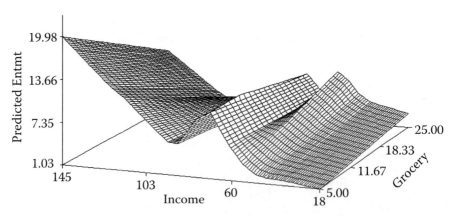

Based on the graph, the analysts arrive at a conclusion that the categories of consumers who spend a lot on entertainment are families with annual income (i) above
$100,000 who spend little on groceries; (ii) between $60,000 and $100,000; and (iii)
under $60,000 per year who spend a lot on groceries.

To predict the amount spent on entertainment by a family with annual income of, say,
$75,000 and grocery expenditure of, say, $10,000, the analysts write the following
code:

```
data point4pred;
 input income grocery;
 cards;
75 10
;

proc loess data=plaza;
 model entmt=income grocery;
   ods output ScoreResults=predicted;
     score data=point4pred;
run;

proc print data=predicted;
run;
```

The relevant output is

```
income    grocery    p_entmt
   75         10      9.81705
```

Thus, the predicted annual amount spent on entertainment by that family is $9,817.05
(roughly, $9,800). □

4.2 Thin-Plate Smoothing Spline Method

4.2.1 Definition

Apart from the loess method of fitting a nonparametric regression, another method may be used. This method is based on the penalized least-squares estimation technique, and is called the *thin-plate smoothing spline* method. It was proposed in a seminal paper in 1976 by a French mathematician Jean Duchon (1949-).[1] The name refers to the fact that when plotted, a smoothing spline surface resembles a bent metal plate. The theoretical foundation of this method follows.

Suppose the response function f is being estimated by the standard least-squares approach in the class of all possible functions. It means that the fitted function is chosen to minimize the residual sum of squares $\sum_{i=1}^{n} \varepsilon_i^2 = \sum_{i=1}^{n} \left(y_i - f(x_{1i}, x_{2i}, \cdots, x_{ki}) \right)^2$, and no restriction is made on the form of the fitted function. In this setting, the function that linearly connects the data points is a possible solution since it reduces this residual sum of squares to zero. It is not a suitable function, however, statistically speaking, because it allows no room for random error. To eliminate solutions of this kind, a restriction on the functional form of the response function must be introduced. This restriction, called the *smoothing penalty*, imposes that f must be an m-times differentiable function over the range of the data, where m is termed the *degree of smoothness* of the function f. Under this restriction, the *penalized least-squares estimator* of f minimizes

$$\frac{1}{n} \sum_{i=1}^{n} \left(y_i - f(x_{1i}, x_{2i}, \cdots, x_{ki}) \right)^2 + \lambda J_m(f)$$

where λ is called the *smoothing parameter*, and

$$J_m(f) = \int_{-\infty}^{\infty} \cdots \int_{-\infty}^{\infty} \sum \frac{m!}{\alpha_1! \alpha_2! \cdots \alpha_k!} \left(\frac{\partial^2 f}{\partial x_1^{\alpha_1} \partial x_2^{\alpha_2} \cdots \partial x_k^{\alpha_k}} \right)^2 dx_1 \cdots dx_k$$

with $\alpha_1 + \alpha_2 + \cdots + \alpha_k = m$. For example, for $m = 2$, the integrand in $J_2(f)$ is the sum of squares of the second-order partial derivatives plus twice the sum of squares of the mixed second-order partial derivatives, that is,

$$J_2(f) = \int_{-\infty}^{\infty} \cdots \int_{-\infty}^{\infty} \left[\left(\frac{\partial^2 f}{\partial x_1^2} \right)^2 + \cdots + \left(\frac{\partial^2 f}{\partial x_k^2} \right)^2 + 2 \left(\frac{\partial^2 f}{\partial x_1 \partial x_2} \right)^2 + \cdots \right.$$
$$\left. + 2 \left(\frac{\partial^2 f}{\partial x_{k-1} \partial x_k} \right)^2 \right] dx_1 \cdots dx_k.$$

For instance, in the case of a single predictor $x_1 = x$,

$$J_2(f) = \int_{-\infty}^{\infty} \left(f''(x) \right)^2 dx,$$

[1]Duchon, J. (1976) Splines minimizing rotation invariant semi-norms in Sobolev spaces, pp 85-100, In: *Constructive Theory of Functions of Several Variables*, Oberwolfach, 1976, W. Schempp and K. Zeller, editors.

and in the case of two predictors x_1 and x_2,

$$J_2(f) = \int_{-\infty}^{\infty} \int_{-\infty}^{\infty} \left[\left(\frac{\partial^2 f}{\partial x_1^2} \right)^2 + 2 \left(\frac{\partial^2 f}{\partial x_1 \partial x_2} \right)^2 + \left(\frac{\partial^2 f}{\partial x_2^2} \right)^2 \right] dx_1 \, dx_2.$$

During the estimation procedure by the method of smoothing splines, the unknown response function f is expanded with respect to a basis in the class of functions with the specified degree of smoothness m, and only the coefficients of this expansion are estimated from the data.

4.2.2   SAS Implementation: Fitting Spline

In SAS, fitting a thin-plate smoothing spline is carried out by running the TPSPLINE procedure. The syntax is

```
PROC TPSPLINE DATA=data_name;
    MODEL response=(predictor1 predictor2 ... predictor_k)/
                M=value ALPHA=value;
        OUTPUT OUT=output_data_name PRED LCLM UCLM;
            SCORE DATA=point4prediction_data_name OUT=predicted_data_name;
RUN;
```

- In the MODEL statement, the list of predictors must be surrounded by parentheses.
- The M= option changes the default value of the degree of smoothness. The default value is max $\left(2, \lceil \frac{k}{2} \rceil \right)$. For example, when $k = 1, 2, 3$ or 4, $M = 2$.
- If the options PRED, LCLM, and UCLM are specified, the output data set *output_data_name* contains columns of values of the predicted response (called P_response), lower confidence limit (called LCLM_response) and upper confidence limit (called UCLM_response) for the 95% confidence interval for predicted response, or the $100(1 - \alpha)\%$ confidence interval, if ALPHA= option is present.
- The SCORE statement specifies a data set with points for which obtaining predicted values is desired. These values are placed in the output data set, and can be viewed by printing this file. The predicted values are in the column called P_response.
- The default value of the smoothing parameter λ is GCV-optimal. The GCV (generalized cross validation) criterion first proposed in 1979[1] minimizes $GCV = \dfrac{n\hat{\sigma}^2}{\left(n - Trace(\mathbb{L})\right)^2}$ (see Subsection 4.1.2 for notation).

4.2.3   SAS Implementation: Plotting Fitted Spline Curve

To plot the fitted spline on a 2D scatterplot, the following code is used. It is similar to the one for plotting a loess curve (see Subsection 4.1.4).

---

[1]Craven, P. and Wahba, G. (1979) Smoothing noisy data with spline functions: Estimating the correct degree of smoothing by the method of generalized cross-validation, *Numerische Mathematik*, **31**, 377-403.

```
SYMBOL1 color=black value=dot color=black;
SYMBOL2 color=black value=none interpol=join line=1;
SYMBOL3 color=black value=none interpol=join line=2;
SYMBOL4 color=black value=none interpol=join line=2;

PROC GPLOT data=output_data_name;
    PLOT (response P_response LCLM_response
                        UCLM_response)*predictor / OVERLAY;
RUN;
```

4.2.4 SAS Implementation: Plotting Fitted Spline Surface

The procedure for plotting the fitted spline surface is exactly the same as the one for plotting the fitted loess surface (cf. Subsection 4.1.6). For convenience we restate it here. First, a data set with the grid points is created using the code:

```
DATA grid_points_data_name;
    DO predictor1=min TO max BY increment;
        DO predictor2=min TO max BY increment;
          OUTPUT;
        END;
    END;
RUN;
```

Next, the procedure TPSPLINE is invoked with the SCORE statement included (see Subsection 4.2.2). The output data set *predicted_data_name* is then used for plotting the 3D scatterplot with the fitted surface. The syntax is

```
PROC G3D data=predicted_data_name;
    PLOT predictor1*predictor2=P_response;
RUN;
```

4.2.5 Examples

Example 4.3 For the data on unemployment rates considered in Example 4.1, we fit and compare two smoothing spline curves with the degrees of smoothness $m = 1$ and 2. Here are the procedures that we run. The degree of smoothness $m = 2$ is the default value in SAS.

```
proc tpspline data=unempl_rates;
   model rate=(month)/m=1;
      output out=result pred lclm uclm;
run;
```

```
title 'm=1';
symbol1 color=black value=dot;
symbol2 color=black value=none interpol=join line=1;
symbol3 color=black value=none interpol=join line=2;
symbol4 color=black value=none interpol=join line=2;

proc gplot data=result;
   plot (rate p_rate lclm_rate uclm_rate)*month /overlay;
run;

proc tpspline data=unempl_rates;
   model rate=(month);
      output out=result pred lclm uclm;
run;

title 'm=2';
proc gplot data=result;
   plot (rate p_rate lclm_rate uclm_rate)*month /overlay;
run;
```

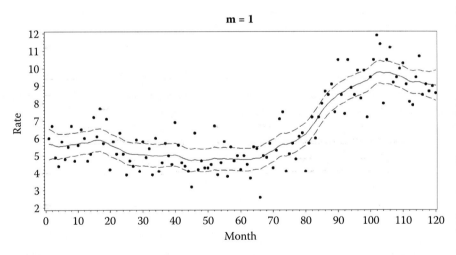

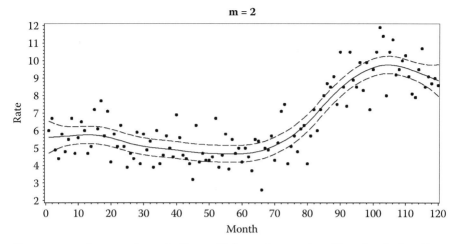

The curve on the second graph (for m=2) is much smoother. Note that both splines don't really do a good job capturing the periodicity in the data. □

Example 4.4 Continuing with Example 4.2, we fit a spline surface on the 3D scatterplot. The required code and the graph are:

```
data grid_points;
   do income=18 to 145 by 1;
       do grocery=5 to 25 by 1;
          output;
       end;
   end;
run;

proc tpspline data=plaza;
   model entmt=(income grocery);
       score data=grid_points out=result;
run;

proc g3d data=result;
   plot income*grocery=p_entmt/ ctop=black cbottom=gray;
run;
```

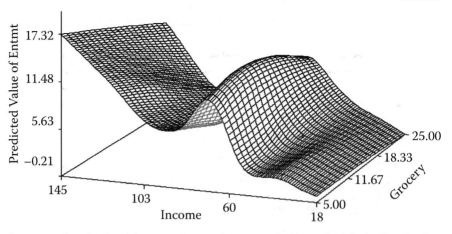

As opposed to the fitted loess surface we have seen in Example 4.2, the fitted spline surface smudges the effect of higher spending on entertainment by consumers who earn less than $60,000 per year and who buy a lot of groceries.

The code below aids in prediction of how much is spent on entertainment by a family that earns $75,000 and spends $10,000 on groceries yearly.

```
data point4pred;
  input income grocery;
  cards;
75 10
;

proc tpspline data=plaza;
  model entmt=(income grocery);
      score data=point4pred out=predicted;
run;

proc print data=predicted;
run;
```

As shown on the output below, the predicted value is $11,174.90 (or about $11,200), which is larger than what was predicted in Example 4.2.

```
income     grocery     P_entmt
   75          10      11.1749
   □
```

Exercises for Chapter 4

Exercise 4.1 The following table contains the population size and total number of crimes in the United States between 1960 and 2010.

Year	Population	Total	Year	Population	Total
1960	179,323,175	3,384,200	1986	240,132,887	13,211,869
1961	182,992,000	3,488,000	1987	242,282,918	13,508,700
1962	185,771,000	3,752,200	1988	245,807,000	13,923,100
1963	188,483,000	4,109,500	1989	248,239,000	14,251,400
1964	191,141,000	4,564,600	1990	248,709,873	14,475,600
1965	193,526,000	4,739,400	1991	252,177,000	14,872,900
1966	195,576,000	5,223,500	1992	255,082,000	14,438,200
1967	197,457,000	5,903,400	1993	257,908,000	14,144,800
1968	199,399,000	6,720,200	1994	260,341,000	13,989,500
1969	201,385,000	7,410,900	1995	262,755,000	13,862,700
1970	203,235,298	8,098,000	1996	265,228,572	13,493,863
1971	206,212,000	8,588,200	1997	267,637,000	13,194,571
1972	208,230,000	8,248,800	1998	270,296,000	12,475,634
1973	209,851,000	8,718,100	1999	272,690,813	11,634,378
1974	211,392,000	10,253,400	2000	281,421,906	11,608,072
1975	213,124,000	11,292,400	2001	285,317,559	11,876,669
1976	214,659,000	11,349,700	2002	287,973,924	11,878,954
1977	216,332,000	10,984,500	2003	290,690,788	11,826,538
1978	218,059,000	11,209,000	2004	293,656,842	11,679,474
1979	220,099,000	12,249,500	2005	296,507,061	11,565,499
1980	225,349,264	13,408,300	2006	299,398,484	11,401,511
1981	229,146,000	13,423,800	2007	301,621,157	11,251,828
1982	231,534,000	12,974,400	2008	304,374,846	11,160,543
1983	233,981,000	12,108,600	2009	307,006,550	10,762,956
1984	236,158,000	11,881,800	2010	308,745,538	10,329,135
1985	238,740,000	12,431,400			

(a) Plot the total number of crimes (in millions) against the population size (in 100 millions). Describe the graph.

(b) Fit an AICC-optimal loess curve using <u>linear</u> local polynomials. Display the curve on the scatterplot along with the 95% confidence band. What is the value of the optimal smoothing parameter? Was the total number of crimes in 2010 below the predicted value? By how much approximately? Was it still within the 95% confidence limits?

(c) Repeat part (b) fitting <u>quadratic</u> local polynomials. Which model, linear or quadratic, has a better fit with respect to the AICC?

Exercise 4.2 A lady has owned a small boutique store for three years. She keeps a record of all her monthly revenues and is interested in finding any discernable patterns. The data are

Year 1	Year 2	Year 3
$6,123.40	$10,063.66	$8,189.96
$7,983.20	$10,372.72	$6,749.13
$8,029.10	$9,462.20	$12,712.92
$8,092.00	$11,316.90	$10,005.40
$6,963.88	$11,724.90	$10,945.87
$6,990.40	$11,318.94	$7,643.48
$9,082.76	$11,978.88	$11,162.35
$9,645.12	$8,943.70	$8,409.41
$8,869.92	$7,547.03	$7,233.91
$9,292.88	$9,878.34	$7,146.34
$6,937.70	$7,430.14	$8,901.20
$10,185.72	$8,062.68	$8,289.41

(a) Fit the loess curve with the optimal value of the smoothing parameter based on the bias corrected Akaike Information Criterion. Fit first degree local polynomials. Construct a 90% confidence band. Display the results on a graph. Describe the pattern you see. Hint: Regress revenue on month. To create a variable for month use the assignment month=_N_.

(b) Fit the loess curve with the optimal value of the smoothing parameter based on the bias corrected Akaike Information Criterion. Fit second degree local polynomials. Construct a 90% confidence band. Display the results on a graph. Describe the pattern you see.

(c) Do the curves constructed in parts (a) and (b) capture periodical (perhaps seasonal?) variations in the data? Plot loess curves for smoothing parameters 0.1, 0.2, 0.3, and 0.4. Fit quadratic local polynomials. Which curve reflects the periodicity the best?

Exercise 4.3 A nutritionist is concerned with her teenage son's eating habits. She surveys his classmates to find out approximately how many cans of soda they drank last week (variable named "Sodas"), and how many servings of French fries (variable named "Fries") and servings of fruits (variable named "Fruits") they ate last week. The data are

Sodas	Fries	Fruits	Sodas	Fries	Fruits
14	8	15	15	6	0
1	1	8	8	5	15
16	6	10	2	2	10
15	6	0	3	3	6
12	6	5	0	0	5
18	7	0	10	2	15
14	7	0	14	2	7
7	5	20	19	7	3
12	6	7	3	2	20
0	0	5	3	3	15
20	7	2	24	7	0
11	1	10	0	1	25
8	6	10	3	4	15
18	6	14	9	3	10
3	3	12	12	3	7

(a) Using the loess method, regress Fruits on Sodas and Fries. Plot a 3D scatterplot and the fitted surface. Describe the pattern of eating habits that you see on the graph.
(b) Last week, the nutritionist's son drank two cans of soda every day, ate French fries twice in a fast food place, and ate an apple and two oranges. Using the loess regression, predict the number of fruits that he was supposed to eat. Compare with reality.

Exercise 4.4 For the data in Exercise 4.1, fit the smoothing spline curves with the degree of smoothness $m = 2, 3$, and 4. Discuss their behavior. Include 95% confidence bands on the graphs. Which spline curve should be preferred?

Exercise 4.5 Refer to Exercise 4.2. Estimate the response function using the thin-plate spline method. Compare results for the degrees of smoothness $m = 2, 3, 4$, and 5. On the scatterplots also display 95% confidence bands. Which spline curve better depicts the periodicity in the data? Which spline curve should be preferred?

Exercise 4.6 Refer to Exercise 4.3.
(a) Plot a fitted spline surface with the degree of smoothness $m = 2$.
(b) Using the fitted spline surface, predict the number of fruits that the nutritionist's son was supposed to eat. Compare it to the reality and to the number predicted in Exercise 4.3.

Exercise 4.7 The TPSPLINE procedure may be used to fit a *semiparametric* model of the form

$$y = \beta_1 z_1 + \cdots + \beta_r z_r + f(x_1, \ldots, x_k) + \varepsilon.$$

Here y is regressed linearly on $z_1, \ldots, z_r$ (thus termed *regression variables*), while all the other predictors, $x_1, \ldots, x_k$, have an unknown relation with y. They are called *smoothing variables*.

The MODEL statement in PROC TPSPLINE should look like this:

MODEL *response* = *list of regression variables* (*list of smoothing variables*);

That is, the list of regression variables precedes the list of smoothing variables which must be put within parentheses.

Consider the data in Exercise 4.3.

(a) Fit a smoothing spline for a semiparametric model with Sodas as a regression variable and Fries as a smoothing variable. Fit the spline surface with the degree of smoothness $m = 2$.

(b) How many fruits should the nutritionist's son have eaten according to this surface? Compare to the reality and the predicted values from Exercises 4.3 and 4.6.

Chapter 5

Nonparametric Generalized Additive Regression

5.1 Definition

A generalized additive model is an extension of a nonparametric linear regression model. Suppose we obtain a sample of n independent observations of k predictor variables $x_1, \ldots, x_k$ and a response y. A *generalized additive regression model* is used to connect through a *link function* $g(\cdot)$ the mean response $\mu = \mathbb{E}(y|x_1, \ldots, x_k)$ and an additive function of the predictors of the form $s_0 + s_1(x_1) + \cdots + s_k(x_k)$ where s_0 is the intercept, and $s_1(\cdot), \ldots, s_k(\cdot)$ are loess or univariate spline smoothers. The equation of the generalized additive model is

$$g(\mu) = s_0 + s_1(x_1) + \cdots + s_k(x_k).$$

In this chapter we focus on two cases: when y is a binary variable and when it is a count variable. With the parametric generalized linear model approach, if y is binary, an ordinary logistic regression model should be fit. If y is a count variable, a Poisson regression model is appropriate. If the linearity of the regression terms may not hold, a nonparametric generalized additive model is a solution.

5.2 Nonparametric Binary Logistic Model

5.2.1 Definition

Suppose the response y assumes only two values 0 and 1. Let $\pi = \mathbb{P}(y = 1)$. Thus, y has a Bernoulli(π) distribution with mean $\mu = \pi$. Define the *logit link function* by

$$g(\pi) = logit(\pi) = \ln\left(\frac{\pi}{1 - \pi}\right).$$

The *nonparametric binary logistic model* has the form

$$g(\pi) = logit(\pi) = \ln\left(\frac{\pi}{1 - \pi}\right) = s_0 + s_1(x_1) + \cdots + s_k(x_k)$$

where s_0 is the intercept, and $s_1(\cdot), \ldots, s_k(\cdot)$ are loess or univariate thin-plate spline smoothers.

This model may be rewritten as

$$\pi = \frac{\exp\{s_0 + s_1(x_1) + \cdots + s_k(x_k)\}}{1 + \exp\{s_0 + s_1(x_1) + \cdots + s_k(x_k)\}},$$

and may be used to estimate the probability of the event $y = 1$ for some particular values of the predictor variables.

Alternatively to a pure nonparametric logistic model, a *semiparametric model* may be considered. In this model, some of the predictors have linear effect, thus only the slopes have to be estimated for these variables. A semiparametric model has the form

$$\pi = \frac{\exp\{s_0 + s_1 x_1 + \cdots + s_i x_i + s_{i+1}(x_{i+1}) + \cdots + s_k(x_k)\}}{1 + \exp\{s_0 + s_1 x_1 + \cdots + s_i x_i + s_{i+1}(x_{i+1}) + \cdots + s_k(x_k)\}}$$

where $s_0, \ldots, s_i$ are constants and $s_{i+1}(\cdot), \ldots, s_k(\cdot)$ are smoothing functions. The predictor variables $x_1, \ldots, x_i$ are called *regression variables*, whereas $x_{i+1}, \ldots, x_k$ are termed *smoothing variables*.

It should be pointed out that SAS distinguishes yet another possible model where two univariate thin-plate splines, say, $s_1(x_1)$ and $s_2(x_2)$ are replaced by a single bivariate spline $s(x_1, x_2)$.

The first use of splines in logistic regression is attributed to American statisticians Trevor John Hastie (1953-) and Robert John Tibshirani (1956-) who in 1986 published a seminal paper on this subject.[1]

5.2.2 SAS Implementation

To fit a pure nonparametric or a semiparametric logistic model, the procedure GAM (stands for "generalized additive models") is called. Three smoothing techniques may be specified: LOESS(·) for loess regression, SPLINE(·) for a *cubic spline* (a univariate thin-plate spline with the degree of smoothness $m = 2$), or SPLINE2(·,·) for a bivariate thin-plate spline. The syntax for fitting a semiparametric model with the loess regression smoothers, for instance, is

```
PROC GAM DATA=data_name;
    CLASS list of categorical predictors;
MODEL response(EVENT='level_name') = PARAM(list of regression predictors)
    LOESS(smoothing_predictor_1) ... LOESS (smoothing_predictor_k)
        /LINK=LOGIST DIST=BINOMIAL;
    OUTPUT PUT=output_data_name PRED;
SCORE DATA=points4prediction_data_name out=predicted_points_data_name;
RUN;
```

---

[1] Hastie, T.J. and Tibshirani, R.J. (1986) Generalized additive models, *Statistical Science*, **1**, 297-318.

• The specification LOESS(*smoothing predictor*) may be replaced by SPLINE(*smoothing predictor*) or SPLINE2(*smoothing_ predictor_1*, *smoothing predictor_2*).

• If smoothing by loess or cubic spline is requested, then instead of the model, say, $logit(\pi) = s_0 + s_1(x_1)$, SAS actually estimates the model $logit(\pi) = s_0 + s_1 x_1 + s_2(x_1)$ where s_0 and s_1 are constants, and $s_2(\cdot)$ is a *de-trended* loess or cubic spline. That is, the procedure GAM separates out the linear trend associated with x_1. If smoothing by the bivariate thin-plate spine is requested, no linear slopes are estimated separately.

• The tests of significance for regression coefficients and linear slopes are conducted based on the *t*-distribution with one degree of freedom. Significance of de-trended loess regression terms is tested based on the chi-square distribution with the degrees of freedom equal to $Trace(\mathbb{L}^T \mathbb{L})$ reduced by one (where $\mathbb{L}$ is the smoothing matrix defined in Subsection 4.1.2). For de-trended cubic splines, inference is done based on the chi-square distribution with three degrees of freedom. Test of significance for bivariate splines is based on the chi-square distribution with four degrees of freedom.

• The option PRED predicts the modeled probability which is saved under the variable name P_response in the data set *output_data_name*. The estimated loess regression (or a de-trended cubic spline) is also stored in that file under the name P_smoothing_predictor. For a bivariate thin-plate spline of predictor1 and predictor2, the estimated values are stored under the name P_predictor1predictor2.

• The SCORE statement computes the fitted response for the points in the input data set and places them into the output data set as the variable P_response.

The syntaxes for plotting 3D scatterplots, 3D fitted surfaces, and multiple graphs are identical to those given in Subsections 4.1.4 – 4.1.6.

5.2.3 Examples

Example 5.1 Peripheral arterial disease (PAD) is a result of plaque buildup in the arteries. In a cardiovascular hospital, investigators would like to regress presence of PAD (1='yes'/0='no') on patient's age (in years) and body mass index (BMI, in kg/m^2). A nonparametric binary logistic regression is fit to the data. SAS is applied with the three different types of smoothers: loess regression, cubic spline, and bivariate spline. The following code fits the models, plots the fitted surfaces, and predicts the probability that a 50-year-old patient with the BMI of 21.4 is diagnosed with PAD.

```
data artery_disease;
input age BMI PAD @@;
datalines;
65 27.3 0   26 20.3 0   46 15.7 0   22 28.5 0   71 19.4 1   33 16.2 1
28 14.3 0   64 20.5 0   73 18.8 1   85 22.4 1   43 26.4 1   51 26.4 0
60 24.3 1   83 18.9 1   26 19.7 0   71 18.7 1   24 21.3 0   71 18.7 1
```

```
48 16.4 1  33 18.8 0  52 21.8 1  70 19.8 0  40 23.4 0  20 22.1 0
85 18.1 1  35 20.6 1  37 18.4 0  54 22.1 0  55 15.5 1  26 19.5 0
33 17.5 0  65 26.2 1  21 21.2 1  76 21.1 1  83 25.5 1  68 20.9 1
31 20.9 0  70 24.8 0  67 21.3 0  34 20.6 1  30 24.4 0  76 19.8 0
26 18.8 0  71 23.2 1  41 25.0 1  73 19.9 1  29 18.7 1  48 23.6 0
34 23.4 0  23 17.5 1  82 16.5 1  23 24.9 0  61 19.9 1  42 18.7 0
47 26.6 1  54 20.2 1  75 14.4 0  52 27.5 1  26 21.3 1  61 21.3 1
51 28.9 0  85 26.7 1  45 26.5 1  77 20.6 0  59 26.9 1  73 22.8 1
20 17.9 0
;
 /* LOESS(age) LOESS(BMI) */

    data point4pred;
    input age BMI;
    cards;
    50 21.4
     ;

data grid_points;
  do age=20 to 85 by 1;
    do BMI=14 to 29 by 1;
output;
    end;
  end;
run;

proc gam data=artery_disease;
    model PAD(event='1')=loess(age) loess(BMI)
              /link=logist dist=binomial;
    output out=result pred;
     score data=point4pred out=predicted;
      score data=grid_points out=score_results;
    run;

title 'LOESS(age) LOESS(BMI)';
proc g3d data=score_results;
  plot age*BMI=P_PAD/ctop=black cbottom=gray name='graph1';
run;

proc print data=predicted;
run;

/*********************************************/

    /*  SPLINE(age) SPLINE(BMI) */
```

```
  proc gam data=artery_disease;
     model PAD(event='1')=spline(age) spline(BMI)
               /link=logist dist=binomial;
     output out=result pred;
      score data=point4pred out=predicted;
       score data=grid_points out=score_results;
     run;

     title 'SPLINE(age) SPLINE(BMI)';
proc g3d data=score_results ;
  plot age*BMI=P_PAD/ctop=black cbottom=gray name='graph2';
run;

proc print data=predicted;
run;

/*************************************************/

   /* SPLINE2(age,BMI) */

proc gam data=artery_disease;
     model PAD(event='1')=spline2(age, BMI)
               /link=logist dist=binomial;
     output out=result pred;
      score data=point4pred out=predicted;
       score data=grid_points out=score_results;
     run;

title 'SPLINE2(age, BMI)';
proc g3d data=score_results;
  plot age*BMI=P_PAD/ctop=black cbottom=gray name='graph3';
run;

proc print data=predicted;
run;

proc greplay igout=work.gseg tc=sashelp.templt template=l2r2
     nofs;
       treplay 1:graph1 2:graph3 3:graph2;
run;
```

The graphs constructed by SAS are:

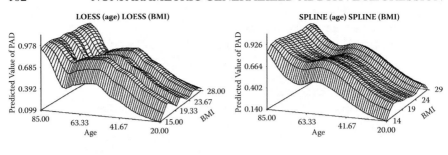

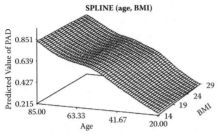

As seen on the graphs, older patients with larger BMI have higher probability of being diagnosed with PAD. Based on the SAS output given below, the predicted probability P_PAD that a 50-year-old patient with the BMI of 21.4 has PAD is 0.69749 for the loess model, 0.61939 for the cubic spline model, and 0.57199 for the bivariate spline model.

```
LOESS(age) LOESS(BMI)
age  BMI   P_PAD
 50 21.4 0.69749

SPLINE(age) SPLINE(BMI)
age  BMI   P_PAD
 50 21.4 0.61939

SPLINE2(age, BMI)
age  BMI   P_PAD
 50 21.4 0.57199
```

☐

Example 5.2 A movie fan collects data about his favorite Hollywood movies. The variables are whether the initial release was a cold opening, that is, no pre-screening for critics was done (yes/no), eventual critics' rating (bad/good), logarithm of production cost and whether the movie was a box office success, that is, whether the total box office revenue was in excess of the production cost (yes/no). The SAS code that reads the instream data is

```
data movies;
input cold_opening $ critic_rating $ log_prod_cost success $ @@;
datalines;
no    bad    13.7  no       no    good   14.1  yes
no    good   14.0  no       no    good   13.1  yes
no    good   13.0  yes      yes   good   14.2  no
yes   good   10.7  no       yes   bad    14.5  no
yes   bad    10.3  no       no    good   14.0  yes
no    good   12.8  no       no    good   14.4  yes
no    bad    12.2  no       no    good   11.5  yes
yes   good   13.0  no       no    bad    13.2  no
no    good   13.4  no       yes   good   12.2  yes
yes   bad    14.0  no       no    good   14.3  yes
no    good   11.6  yes      no    good   14.5  yes
no    good   9.7   no       no    good   14.0  yes
no    good   11.1  yes      no    good   13.0  yes
no    good   14.2  yes      no    good   13.7  yes
no    good   14.1  no       no    bad    13.3  no
no    good   12.5  yes      no    good   14.0  yes
no    good   13.4  yes      no    good   12.1  yes
no    good   14.2  no       no    good   14.1  no
no    good   13.0  yes      no    bad    14.2  no
no    good   14.2  yes      no    good   13.5  yes
yes   bad    13.9  yes      no    good   14.4  no
no    good   14.2  yes      no    good   17.6  no
no    good   14.3  yes      no    good   14.0  yes
yes   good   12.5  yes      yes   good   13.1  no
no    good   13.5  no       yes   good   13.4  no
```

Since some of the predictor variables are binary, we fit the semiparametric logistic model of the form

$$logit\,\mathbb{P}(\text{success}) = s_0 + s_1\,\text{cold\_opening} + s_2\text{critic\_rating}$$
$$+ s_3(\text{log\_prod\_cost})$$

where s_0, s_1, and s_2 are constants, and $s_3(\cdot)$ is a cubic smoothing spline. The code is

```
proc gam data=movies;
class cold_opening critic_rating;
model success(event='yes')=param(cold_opening critic_rating)
              spline(log_prod_cost)/link=logist dist=binomial;
   output out=result pred;
run;

proc print data=result;
run;
```

The estimated coefficients and a partial output for the estimated de-trended spline are

```
                          Parameter
Parameter                 Estimate  Pr > |t|
Intercept                  1.92477    0.6892
cold_opening  no           1.43576    0.1179
critic_rating bad         -2.64060    0.0303
Linear(log_prod_cost)     -0.18174    0.6228
```

```
Source                     Pr > ChiSq
Spline(log_prod_cost)         0.0465
```

Obs	cold_opening	critic_rating	log_prod_cost	success	P_success	P_log_prod_cost
1	no	bad	13.7	no	0.12829	-0.14629
2	no	good	14.1	yes	0.69048	0.00433
3	no	good	14.0	no	0.68951	-0.01836
4	no	good	13.1	yes	0.70371	-0.11472
5	no	good	13.0	yes	0.72571	-0.02495
		---	lines omitted	---		
46	no	good	14.0	yes	0.68951	-0.01836
47	yes	good	12.5	yes	0.56461	0.60687
48	yes	good	13.1	no	0.36107	-0.11472
49	no	good	13.5	no	0.66292	-0.23072
50	yes	good	13.4	no	0.31989	-0.24377

Note that the eventual critics' rating and the de-trended spline are significant at the 5% level, while the indicator of cold opening and linear trend in log production cost are not significant predictors. The fitted model is

$$logit(\text{P\_success}) = 1.92477 + 1.43576 * \text{no\_cold\_opening}$$

$$-2.64060 * \text{bad\_critic\_rating} - 0.18174 * \text{log\_prod\_cost} + \text{P\_log\_prod\_cost}.$$

Next, we plot the estimated probability of success against logarithm of production cost for all level combinations of indicators of cold opening and eventual critics' rating. The code is

```
proc sort data=result;
  by cold_opening critic_rating log_prod_cost;
run;

symbol1 color=black value=none interpol=join line=1;
proc gplot data=result;
    plot p_success*log_prod_cost/name='graph';
    by cold_opening critic_rating;
```

```
    run;

proc greplay igout=work.gseg tc=sashelp.templt template=l2r2
    nofs;
        treplay 1:graph 2:graph2 3:graph1 4:graph3;
run;
```

The graphs are

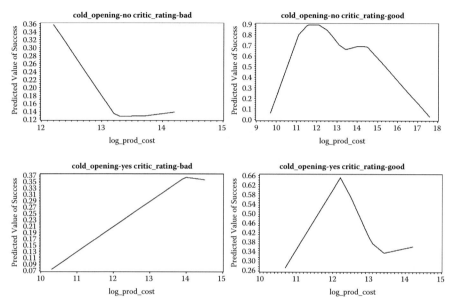

From the graphs, the movies that have a better chance of making money at the box office are: (i) low-budget bad movies pre-screened for critics; (ii) high-budget non-pre-screened bad movies; and (iii) mid-budget good movies.

Finally, we ask SAS to estimate the probability that, for instance, a pre-screened good movie with log-production cost of 10.0 is a box office success. The statements are

```
data point4pred;
input cold_opening $ critic_rating $ log_prod_cost;
cards;
no good 10.0
;

proc gam data=movies;
class cold_opening critic_rating;
    model success(event='yes')=param(cold_opening critic_rating)
        spline(log_prod_cost)/link=logist dist=binomial;
            score data=point4pred out=predicted;
```

```
run;

proc print data=predicted;
run;
```
The estimated probability is 0.14743 as shown in the output.

		log_	
cold_	critic_	prod_	
opening	rating	cost	P_success
no	good	10	0.14743

☐

5.3   Nonparametric Poisson Model

5.3.1   Definition

Suppose the response variable Y assumes integer values 0, 1, 2, etc. This type of observation is called *count data*. The frequencies of counts follow a Poisson distribution if the following features are present:

- Sample mean is approximately equal to sample variance.
- Smaller values are the most common observations.
- Large observations are very infrequent.
- Not too many zeros are observed.

When the frequencies of counts follow a Poisson distribution, the Poisson regression may be used to model the data. The *Poisson regression model* specifies that the response variable Y, given predictors $X_1, \ldots, X_k$, follows a Poisson distribution with the probability mass function

$$\mathbb{P}(Y = y | X_1 = x_1, \ldots, x_k = x_k) = \frac{\lambda^y}{y!} e^{-\lambda}, \ \ y = 0, 1, 2, \ldots,$$

where, in the case of
- parametric regression, the rate λ satisfies

$$\ln \lambda = \ln \mathbb{E}(Y | x_1, \ldots, x_k) = \beta_0 + \beta_1 x_1 + \cdots + \beta_k x_k.$$

- nonparametric additive regression,

$$\ln \lambda = s_0 + s_1(x_1) + \cdots + s_k(x_k)$$

where s_0 is a constant and $s_1(\cdot), \ldots, s_k(\cdot)$ are loess regression or cubic spline smoothers.

• semiparametric additive regression,

$$\ln \lambda = s_0 + s_1 x_1 + \cdots + s_i x_i + s_{i+1}(x_{i+1}) + \cdots + s_k(x_k)$$

where $s_0, \ldots, s_i$ are constants, and $s_{i+1}(\cdot), \ldots, s_k(\cdot)$ are smoothers.

Splines were first applied to model the rate in a Poisson regression in 1986 by an Irish statistician Finbarr O'Sullivan (1958-), and American statisticians Brian S. Yandell (1952-), and William J. Raynor Jr. (1949-).[1]

5.3.2   SAS Implementation

The syntax for fitting a nonparametric (or semiparametric) Poisson model is similar to the one for the nonparametric (or semiparametric) logistic model with adjusted specifications for the link function and the probability distribution. The syntax is

```
PROC GAM DATA=data_name;
    CLASS list of categorical predictors;
MODEL response = PARAM(list of regression predictors)
    LOESS(smoothing_predictor_1) ... LOESS (smoothing_predictor_k)
        /LINK=LOG DIST=POISSON;
    OUTPUT PUT=output_data_name PRED;
SCORE DATA=points4prediction_data_name out=predicted_points_data_name;
RUN;
```

• The specification LOESS(smoothing_predictor) may be replaced by SPLINE(smoothing_predictor) or SPLINE2(smoothing_predictor_1, smoothing_predictor_2).
• All graphics related syntaxes may be found in Subsections 4.1.4 – 4.1.6.

5.3.3   Examples

Example 5.3 Study participants were asked their age (in years), general health status (poor, fair, good, or excellent), and the number of times they visited a doctor last month. The code below fits a semiparametric Poisson regression model with the rate λ satisfying

$$\ln \lambda = s_0 + s_1 \, \texttt{poor} + s_2 \, \texttt{fair} + s_3 \, \texttt{good} + s_4(\texttt{age})$$

where $s_0, \ldots, s_3$ are constants and $s_4(\cdot)$ is a cubic spline.

```
data docvisits;
  input health $ age n_visits @@;
    datalines;
```

---

[1]O'Sullivan, F., Yandell, B.J. and Raynor,W.J. Jr. (1986) Automatic smoothing of regression functions in generalized linear models. *Journal of the American Statistical Association*, **81**, 96-103.

```
exclnt 18  0   exclnt 53  1   good   39  3   good   59  1
good   24  0   exclnt 39  1   good   51  3   good   57  1
exclnt 23  0   good   42  1   fair   56  4   fair   56  1
exclnt 48  0   exclnt 43  1   good   66  4   exclnt 38  1
exclnt 23  0   good   52  1   good   59  4   good   38  1
exclnt 28  0   fair   37  3   fair   67  4   good   41  1
good   54  0   exclnt 54  3   fair   68  4   fair   40  1
good   29  0   exclnt 40  3   fair   69  4   exclnt 36  1
good   35  0   good   56  3   fair   58  4   good   42  1
good   44  0   good   67  3   fair   57  4   fair   52  1
fair   56  0   fair   47  3   fair   54  2   good   41  1
good   33  1   good   41  3   exclnt 44  2   good   27  1
good   52  1   fair   41  3   good   57  2   fair   25  2
good   50  1   good   39  3   good   57  2   good   44  2
fair   49  1   good   52  3   fair   59  2   fair   50  2
fair   48  1   fair   47  3   good   53  2   good   32  2
exclnt 22  1   fair   70  3   good   52  2   good   53  2
exclnt 39  1   good   62  3   exclnt 43  2   exclnt 56  2
exclnt 54  1   good   50  3   good   28  2   fair   45  2
good   40  1   good   52  3   good   52  2   good   50  2
good   47  1   good   68  3   fair   60  2   fair   49  2
exclnt 44  1   good   57  3   exclnt 19  2   good   21  2
fair   55  1   fair   58  3   good   60  2   good   44  2
good   43  1   good   49  3   fair   33  2   good   55  2
fair   50  1   fair   68  3   good   43  3   exclnt 36  2
good   72  4   fair   54  3   good   48  2   poor   77  6
good   67  4   good   66  3   good   48  4   poor   74  6
fair   78  5   good   33  3   good   60  4   fair   58  6
fair   70  5   exclnt 30  3   good   66  4   good   68  7
poor   72  5   good   62  3   good   66  4   poor   68  7
poor   67  5   good   60  3   fair   66  4   poor   71  8
fair   74  5   fair   64  3   good   65  4   fair   55  2
fair   54  5   good   38  3   fair   69  4   good   58  2
poor   62  5   exclnt 66  3   good   58  4   good   49  2
good   73  5   exclnt 28  3   good   69  4   exclnt 24  2
good   55  2   poor   69  5   fair   64  4   good   44  3
exclnt 49  2   fair   69  6   fair   45  2   poor   77  6
good   46  2   poor   75  6
;

proc sort data=docvisits;
by descending health; /* To make "excellent" a reference */
run;

proc gam data=docvisits;
```

```
class health (order=data);
  model n_visits=param(health) spline(age)/link=log
                  dist=poisson;
  output out=result pred;
run;

proc print data=result;
run;
```

A partial output is

	Parameter	
Parameter	Estimate	Pr > |t|
Intercept	-0.72057	0.0051
health poor	0.55264	0.0294
health good	0.19599	0.3007
health fair	0.27289	0.1820
Linear(age)	0.02663	<.0001

Source	Pr > ChiSq
Spline(age)	0.2519

Obs	health	age	n_visits	P_n_ visits	P_age
1	poor	69	5	5.68308	0.06781
2	poor	75	6	6.41772	0.02959
3	poor	77	6	6.61782	0.00703
4	poor	77	6	6.61782	0.00703
5	poor	74	6	6.31459	0.04002
--- lines omitted ---
146	exclnt	44	1	1.51304	-0.03712
147	exclnt	19	2	0.94137	0.15414
148	exclnt	66	3	2.97664	0.05365
149	exclnt	36	2	1.41614	0.10975
150	exclnt	28	3	1.21322	0.16815

The estimated rate $\widehat{\lambda}$ has the form

$$\ln\widehat{\lambda} = -0.72057 + 0.55264\text{poor} + 0.27289\text{fair} + 0.19599\text{good}$$
$$+0.02663\text{age} + \text{P\_age}.$$

Note that in this model only the intercept, health status "poor," and linear trend of age are significant predictors at the 5% level.

To predict the number of doctor visits for, say, a 70-year-old person in good health (someone not in the original data set), we run the following code:

```
data point4pred;
  input health $ age;
 cards;
 good 70
 ;

proc gam data=docvisits;
  class health(order=data);
    model n_visits=param(health) spline(age)/link=log
                  dist=poisson;
    score data=point4pred out=predicted;
run;

proc print data=predicted;
run;

health age P_n_visits
  good  70    4.08142
```

Thus, roughly 4 visits are predicted.

Next, we plot the fitted number of doctor visits against age stratified by health status. The segment of code and the displayed graphs are below.

```
proc sort data=result;
 by health age;
run;

symbol1 color=black value=none interpol=join line=1;

proc gplot data=result;
  plot p_n_visits*age /name='graph';
    by health;
run;

proc greplay igout=work.gseg tc=sashelp.templt template=l2r2
 nofs;
    treplay 1:graph 2:graph2 3:graph1 4:graph3;
run;
```

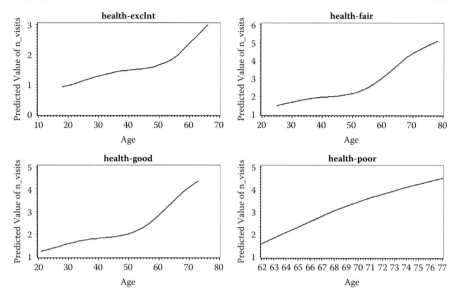

These graphs support the intuition in the sense that older people, especially those in poor health, have more doctor visits. □

Example 5.4 Meadow vole, also known as field mouse of North America, is a small rodent that can do unbelievable damage to agricultural crops. Agronomists study the relation between the lowest winter temperature (in degrees Fahrenheit), the total rainfall in the spring (in inches), and the number of field mice per acre in mid-summer. They collect the data for 20 regions. The following code fits a nonparametric Poisson model to the data.

```
data field_mice;
input temp rainfall n_mice @@;
datalines;
-10 36.4 1    10 27.1 0
-20 43.3 0    -9 32.9 1
 14 25.5 2   -16 23.7 2
 24 22.0 2   -15 22.4 0
-15 17.2 4    25 31.8 4
 15 32.3 4     7 23.1 5
 28 38.0 1     4 29.2 6
  0 28.1 6     5 14.2 9
  7 10.4 8    -4 24.6 3
  0 27.7 8    -8 19.1 2
;

proc gam data=field_mice;
    model n_mice=spline(temp) spline(rainfall)/link=log
```

```
                dist=poisson;
    output out=result pred;
run;

proc print data=result;
run;
```

SAS outputs the fitted parameters and the estimated spline.

	Parameter	
Parameter	Estimate	Pr > |t|
Intercept	2.33372	<.0001
Linear(temp)	0.00825	0.4978
Linear(rainfall)	-0.04015	0.0295

Source	Pr > ChiSq
Spline(temp)	0.0413
Spline(rainfall)	0.2537

Obs	temp	rainfall	n_mice	P_n_mice	P_temp	P_rainfall
1	-10	36.4	1	1.14397	-0.45938	-0.19567
2	10	27.1	0	3.15302	-0.17688	-0.00281
3	-20	43.3	0	0.15952	-0.88315	-1.38237
4	-9	32.9	1	2.10108	-0.38721	0.19132
5	14	25.5	2	2.53926	-0.35816	-0.13527
		--- lines omitted ---				
16	5	14.2	9	8.79123	0.28791	0.08106
17	7	10.4	8	8.37229	0.10456	0.04649
18	-4	24.6	3	3.48072	0.12315	-0.18880
19	0	27.7	8	5.55330	0.43854	0.05442
20	-8	19.1	2	2.89867	-0.30117	-0.13531

Note that the slope for rainfall and spline for temperature are significant at the 5% level, whereas the slope for temperature and spline for rainfall are insignificant. The fitted mean number of field mice per acre is expressed as

$$\widehat{\lambda} = \exp\{2.33372 + 0.00825\text{temp} - 0.04015\text{rainfall} + \text{P\_temp} + \text{P\_rainfall}\}.$$

The 3D scatterplot for the predicted number of field mice against temperature and rainfall is plotted using the code:

```
proc g3d data=result;
  scatter temp*rainfall=p_n_mice/color="black";
run;
```

The plot is

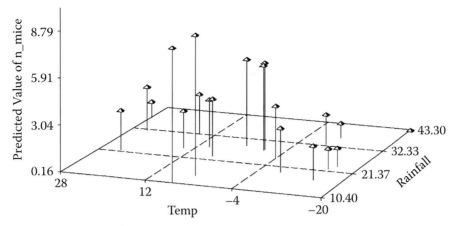

As seen on this graph, colder winters, or warmer winters with more spring rainfall result in a reduced number of field mice. □

Exercises for Chapter 5

Exercise 5.1 Sport betting is very popular in men's basketball in the National Collegiate Athletic Association (NCAA). Each season, conferences (associations of schools) hold tournaments with home-and-home round robin schedules, where each pair of teams in the conference plays one game on each other's home court for a total of two games. The margin of victory (or loss) for a team is defined as the point differential at the end of the game.

The data below are collected on 150 pairs of teams from different conferences. The data consist of the margin of victory (or loss) for the first home team, and whether it won (yes/no) on the road (that is, on the opponent's home court).

\multicolumn{10}{c}{Margin at Home and Whether Won on the Road}
3	yes	13	yes	3	yes	15	yes	2	no	16	yes
5	no	21	yes	−8	yes	6	yes	−7	yes	5	yes
9	yes	16	yes	−12	no	−1	yes	1	yes	7	no
10	yes	−6	yes	−1	yes	−1	no	3	yes	−21	yes
4	no	6	yes	25	no	6	no	6	yes	6	yes
30	yes	−1	yes	−23	no	1	yes	−2	yes	9	yes
4	no	29	yes	14	no	6	no	21	no	8	yes
−2	no	7	no	2	yes	11	yes	5	no	6	yes
−14	no	6	no	2	yes	3	yes	7	yes	3	yes
−8	yes	5	yes	9	no	-12	yes	3	yes	8	yes
2	no	7	yes	−5	no	2	yes	−16	yes	7	no
4	no	3	no	2	no	5	no	−16	no	1	no
−23	no	−14	no	14	yes	−13	no	1	yes	5	no
−16	no	14	yes	7	yes	1	yes	1	no	13	yes
−1	no	6	yes	5	yes	−2	yes	−3	yes	−7	no
17	yes	3	yes	8	no	1	no	9	no	−4	no
7	no	5	yes	9	yes	13	yes	−15	no	8	yes
23	yes	2	yes	1	no	−22	yes	−1	yes	10	no
−5	no	4	no	−12	no	−4	yes	8	yes	6	no
25	yes	−4	no	9	yes	11	yes	8	yes	12	yes
9	yes	2	yes	27	yes	−15	yes	5	yes	8	yes
10	yes	−9	no	−4	no	5	yes	3	no	9	yes
6	no	4	yes	−16	yes	7	yes	2	yes	−10	yes
17	yes	4	no	16	yes	7	yes	2	yes	−8	no
−23	no	7	yes	22	yes	14	no	−22	no	9	no

(a) Fit a nonparametric logistic model with the loess regression smoother. Is the margin of victory a significant predictor of winning on the road? Write down the fitted regression model with estimated coefficients.

(b) Predict the probability of winning on the road for a team that lost by 11 points on its court.

(c) Construct a graph showing the predicted probability of winning as a visiting team against the margin of victory as a home team.

(d) Redo parts (a)-(b), smoothing with a cubic spline.

Exercise 5.2 Adherence to medication regimen is poor in pediatric patients with chronic illness. A study is conducted that is aimed at examining effectiveness of a token system intervention. Children in the intervention group receive tokens that can be exchanged for prizes. Gender (M/F), intervention (yes/no), age (in years), and adherence to medication (yes/no) are recorded for each study participant. The data are

Gender	Interv.	Age	Adher.	Gender	Interv.	Age	Adher.
F	no	10	yes	F	yes	8	yes
F	no	9	yes	F	yes	10	yes
M	yes	8	yes	M	yes	11	yes
M	yes	11	yes	F	yes	8	yes
F	yes	10	yes	M	yes	6	no
F	yes	7	no	F	yes	12	yes
M	no	6	no	M	no	7	no
M	no	6	no	F	no	12	no
M	yes	7	no	F	no	11	no
M	yes	12	yes	M	yes	6	no
M	yes	10	yes	F	no	12	no
F	yes	8	yes	F	no	11	no
M	no	7	no	F	yes	10	yes
F	yes	6	yes	M	no	11	no
M	no	12	no	M	no	9	no
F	yes	7	yes	M	no	10	no
M	no	8	yes	M	yes	6	yes
M	no	6	no	F	no	10	no
F	no	8	no	M	yes	6	no
M	yes	10	yes	M	yes	8	yes

(a) Fit a binary logistic model of the form

$$logit\,\mathbb{P}(\text{Adherence} =' \text{yes}') = s_0 + s_1\,\text{Gender} + s_2\text{Intervention} + s_3(\text{Age})$$

where s_0, s_1, and s_2 are constants, and $s_3(\cdot)$ is a cubic spline. Which variables are significant predictors of medication adherence? Write down the estimated model.
(b) Plot the predicted probability of adherence against age by gender and intervention (a total of four graphs). Discuss the patterns you see on the graphs.

Exercise 5.3 The U.S. Department of Veterans Affairs offers a variety of programs to treat posttraumatic stress disorder (PTSD) following combat exposure in veterans. The PTSD symptom severity score (ranging between 17 and 85) is a total of 17 self-reported items each measured on a five-point scale. If this score is at least 50, the patient is diagnosed with PTSD. Data on current age (in years), gender, combat injury (yes/no), and deployment length (in month) are obtained by investigators for 52 veterans. The data are

Age	Gender	Injury	Depl.	Score	Age	Gender	Injury	Depl.	Score
41	M	yes	24	66	23	F	yes	16	60
24	M	yes	19	56	40	M	yes	10	78
36	M	no	4	19	29	M	yes	21	59
41	M	yes	27	79	33	M	yes	23	65
26	M	yes	18	83	39	M	yes	4	46
40	F	yes	27	82	43	M	no	16	42
29	M	yes	15	75	31	M	yes	18	65
40	M	yes	8	26	36	M	yes	27	70
34	M	yes	16	67	23	M	no	22	34
30	M	no	9	29	42	M	yes	7	32
38	F	no	10	64	44	F	no	9	27
31	F	yes	4	78	45	F	no	21	58
47	M	yes	20	60	39	M	yes	25	63
40	M	no	18	63	42	M	yes	8	67
24	F	yes	8	55	36	M	yes	26	70
26	M	yes	10	68	42	M	yes	12	67
42	M	no	6	32	40	M	yes	25	80
40	M	yes	14	70	33	F	no	23	41
24	M	no	13	29	25	F	yes	21	70
25	M	yes	17	68	27	M	yes	21	62
27	M	no	14	43	40	M	yes	7	22
39	F	yes	22	73	24	M	yes	16	59
42	M	no	19	40	46	M	yes	4	57
39	M	yes	20	83	43	M	no	22	42
43	M	yes	6	67	27	M	yes	9	62
34	M	yes	12	79	31	F	no	9	34

(a) Develop a model for the probability of PTSD diagnosis with gender and combat injury as regression variables (since they are binary), and age and deployment length smoothed by a bivariate thin-plate spline. What regressors are significant at the 5% level? Write down the fitted model explicitly. Predict the probability of PTSD presence in a 25-year-old injured male veteran who has been deployed for 12 months.
(b) Develop a model as in part (a) but use a sum of two univariate cubic splines. Write down the fitted model and comment on significance of its components. Repeat the prediction and compare the value to what was obtained in part (a).

Exercise 5.4 A statistics student from a commuter university is interested in verifying the hypothesis that there are more traffic accidents eastbound in the morning and westbound in the evening, because cars are going towards the sun. He collects ten days' worth of data on the number of accidents on the stretch of the freeway that he commutes on Monday through Friday. The variables are the time of the day plus traffic bound (morning east/morning west/afternoon east/afternoon west), the brightness of the sunlight (on a scale 0 to 10), and the number of reported accidents. The data are

Time+Bound	Sunlight Brightness	Number of Accidents	Time+Bound	Sunlight Brightness	Number of Accidents
morning_east	5	4	morning_east	6	3
morning_west	5	1	morning_west	6	0
afternoon_east	4	3	afternoon_east	5	1
afternoon_west	4	5	afternoon_west	5	3
morning_east	2	2	morning_east	7	4
morning_west	2	0	morning_west	7	3
afternoon_east	5	3	afternoon_east	4	3
afternoon_west	5	6	afternoon_west	4	2
morning_east	6	1	morning_east	7	2
morning_west	6	1	morning_west	7	0
afternoon_east	4	4	afternoon_east	5	1
afternoon_west	4	4	afternoon_west	5	3
morning_east	8	5	morning_east	8	5
morning_west	8	3	morning_west	8	4
afternoon_east	3	2	afternoon_east	5	3
afternoon_west	3	3	afternoon_west	5	4
morning_east	4	5	morning_east	4	4
morning_west	4	3	morning_west	4	3
afternoon_east	5	4	afternoon_east	4	1
afternoon_west	5	8	afternoon_west	4	5

(a) Fit a nonparametric Poisson regression to model the number of accidents. Use the loess regression smoother. Write down the estimated model. What significance is present in the model? Does the model support the hypothesis of interest? Explain.

(b) Repeat part (a) but use a cubic spline component.

Exercise 5.5 An avid newspaper reader who had a lot of free time on his hands, counted the number of typos on the front page of 28 issues of different newspapers. For each newspaper, he recorded the circulation ($\times 10,000$ copies) and cost (in US dollars). The data are

Circulation	Cost	Number of Typos	Circulation	Cost	Number of Typos
10.0	2.80	2	45.0	4.50	6
9.0	2.99	2	10.5	0.99	4
1.5	0.25	7	13.5	3.45	0
15.0	3.85	1	13.0	2.99	2
40.0	3.5	1	1.5	0.45	7
2.5	0.99	5	2.0	1.60	6
6.5	1.7	2	3.5	0.75	8
8.5	1.75	4	18.5	3.0	2
1.5	0.0	7	23.5	2.0	3
0.5	0.0	8	4.5	3.8	3
10.5	5.45	0	14.5	3.6	2
25.5	3.45	0	22.0	2.5	1
4.5	1.5	3	11.5	1.65	2
2.0	1.0	6	44.5	2.45	5

(a) Fit a nonparametric Poisson regression smoothed by univariate cubic splines to model the number of typos. Write down the theoretical model. What model is actually estimated by SAS? Write down the fitted model explicitly.

(b) If 13,500 copies of a newspaper are printed and sold at $1.75 each, what is the estimated number of typos?

(c) Plot the three-dimensional fitted surface. Comment.

(d) Redo parts (a)-(c) for the model with bivariate thin-plate smoothing spline.

Chapter 6

Time-to-Event Analysis

Frequently investigators are interested in modeling the distribution of the time that elapses before a certain event occurs. Generally speaking, the modeling of the distribution is called the *time-to-event analysis* but because of the wide range of applications, this analysis may bear other names depending on the field. For instance, in a clinical trial the event may be a medical complication in a patient or even death, and the analysis is called the *survival analysis*. In industrial processes, a machinery component malfunction may be the event of interest, and the analysis is termed *reliability analysis* or *time-to-failure analysis* or *duration analysis*.

The distinctive feature of the time-to-event analysis is modeling in the presence of censored observations. A time is called *censored* if no event has occurred up to this time but no further information on the status of the individual is known. For instance, a patient in a drug trial is lost to follow-up visits due to geographical relocation. Another example of censoring is when at the end of a study some individuals are still event-free.

Traditionally, the distribution of time to event is modeled not in terms of cumulative distribution function but one minus it. It represents the probability that the event occurs after a certain time. This function is termed *reliability function* in engineering, but in a broader range of applications it is called the *survival function* or *survivor function*. In this chapter the term *survival function* will be used.

Several parametric models for survival function have been proposed. Exponential, Weibull, log-normal are some of them. In this chapter we study the nonparametric techniques that are commonly applied: the Kaplan-Meier estimator of the survival function, the log-rank test for comparison of two survival functions, and the Cox proportional hazards model.

6.1 Kaplan-Meier Estimator of Survival Function

Denote by T the time to an event. Let $F_T(t) = \mathbb{P}(T \leq t)$ denote its cumulative distribution function. The *survival function* of T is defined as the probability to survive longer than time t, that is, $S_T(t) = 1 - F_T(t) = \mathbb{P}(T > t)$.

Suppose $t_1 < t_2 < \cdots < t_k$ are k distinct ordered times to event. Note that there might be ties in the data: two or more events can occur at the same time. Also, along with the event, a censoring can occur at some of these times. Denote by $n_i, i = 1, \ldots, k$, the number of individuals *at-risk* at time t_i, that is, those who still haven't experienced the event shortly before this time, and let e_i be the number of individuals who experienced the event at time t_i. The *Kaplan-Meier* (*KM* or *product-limit*) estimator of the survival function is

$$\widehat{S}(t) = \prod_{i:t_i \leq t} \left(1 - \frac{e_i}{n_i}\right), \quad t \geq 0. \tag{6.1}$$

This estimator was developed in 1958 as a result of collaboration between two Americans, a mathematician Edward L. Kaplan (1920-2006) and a biostatistician Paul Meier (1924-2011).[1]

The plot of the KM estimator against time is called the *Kaplan-Meier survival curve*. It is a step function with vertical lines corresponding to the event times. The curve is tick-marked by some symbol at the censoring times (for example, a circle or an "x"). If censoring occurs at an event time, it is conventional to put the symbol at the bottom of the step.

6.1.1 Derivation of KM Estimator

Denote by $\pi_i = \mathbb{P}(T > t_i \mid T > t_{i-1})$, the conditional probability that no event has happened before and including time t_i, given that no event has occurred on or before the preceding event time t_{i-1}, $i = 1, \ldots, k$. For convenience of notation, we assume that $t_0 = 0$.

The survival function at some fixed event time t_j may be written recursively as

$$S(t_j) = \mathbb{P}(T > t_j) = \mathbb{P}(T > t_j \mid T > t_{j-1})\mathbb{P}(T > t_{j-1})$$
$$= \pi_j S(t_{j-1}) = \cdots = \prod_{i=1}^{j} \pi_i. \tag{6.2}$$

The probabilities π_i's are estimated by the method of maximum likelihood. At any event time t_i, there are e_i individuals who experience the event at that time with probability $1 - \pi_i$ each, independently of all others, and there are $n_i - e_i$ individuals who are at risk but don't experience the event with probability π_i each, also independently of all others. Thus, the likelihood function has the form

$$L(\pi_1, \ldots, \pi_k) = \prod_{i=1}^{k} (1 - \pi_i)^{e_i} \pi_i^{n_i - e_i}.$$

[1] Kaplan, E. L. and Meier, P. (1958) Nonparametric estimation from incomplete observations, *Journal of the American Statistical Association*, **53**, 457-481.

Setting to zero the partial derivatives of the log-likelihood function $\ln L = \sum_{i=1}^{k} \left[e_i \ln(1 - \pi_i) + (n_i - e_i) \ln \pi_i \right]$, we arrive at the system of normal equations

$$0 = \frac{\partial \ln L}{\partial \pi_i} = -\frac{e_i}{1 - \pi_i} + \frac{n_i - e_i}{\pi_i}, \quad i = 1, \ldots, k.$$

Thus, the maximum likelihood estimator $\widehat{\pi}_i$ of π_i solves

$$\frac{e_i}{1 - \widehat{\pi}_i} = \frac{n_i - e_i}{\widehat{\pi}_i}, \quad i = 1, \ldots, k.$$

The solution is

$$\widehat{\pi}_i = 1 - \frac{e_i}{n_i}, \quad i = 1, \ldots, k.$$

Substituting this expression into (6.2), we obtain that at any event point t_j,

$$\widehat{S}(t_j) = \prod_{i=1}^{j} \left(1 - \frac{e_i}{n_i} \right).$$

For any time t such that $t_j < t < t_{j+1}$, the estimator of the survival function at that point, $\widehat{S}(t)$, coincides with $\widehat{S}(t_j)$ since no events occur between times t_j and t. This results in (6.1).

6.1.2 SAS Implementation

PROC LIFETEST computes the Kaplan-Meier estimator of the survival function and plots the survival curve. It is a default method of estimation used by SAS. For each individual, the data set must contain the observed event or censoring time and an indicator of whether a censoring has occurred at that time. The syntax is

```
PROC LIFETEST DATA=data_name plots=(survival);
    TIME time_variable_name * censoring_variable_name(1);
RUN;
```

• The digit 1 in parentheses, (1), symbolizes that *censoring_variable_name*=1 for censored times, and 0 otherwise.

6.1.3 Example

Example 6.1 Pilot training school officials are not satisfied with knowing the average number of flight hours it takes their students to obtain a private pilot certificate. They realize that there is a great deal of variability in the number of hours, and, moreover, a substantial percentage of those who originally enroll drop out of the program. They collect the data to conduct the time-to-event analysis. The observations in increasing order are

2+ 5+ 8+ 8+ 10+ 16+ 22+ 36+ 40 42+ 47 50 50+ 52 52 52 59 67 84+

Here some of the enrollees dropped out and some are still in training at the time the data are collected. Their times are censored. Traditionally, censored observations are marked by a plus sign.

The six distinct times to obtaining the certification are 40, 47, 50, 52, 59, and 67 hours. The KM estimation is best done by means of the following table.

Time, t_i	At Risk, n_i	Experienced Event, d_i	Censored at t_i	Survival Rate, $1 - d_i/n_i$	Estimator $\widehat{S}(t)$, $t_i \leq t < t_{i+1}$
0	19	0	0	$1 - 0 = 1.00$	1.00
40	11	1	0	$1 - \frac{1}{11} = 0.91$	(1.00)(0.91)=0.91
47	9	1	0	$1 - \frac{1}{9} = 0.89$	(0.91)(0.89)=0.81
50	8	1	1	$1 - \frac{1}{8} = 0.88$	(0.81)(0.88)=0.71
52	6	3	0	$1 - \frac{3}{6} = 0.50$	(0.71)(0.50)=0.35
59	3	1	0	$1 - \frac{1}{3} = 0.67$	(0.35)(0.67)=0.24
67	2	1	0	$1 - \frac{1}{2} = 0.50$	(0.24)(0.50)=0.12

We can check that our estimates are correct by running the following SAS code:

```
data pilot_certificate;
    input hours censored @@;
  datalines;
 2 1   5 1   8 1    8 1
10 1  16 1  22 1   36 1
40 0  42 1  47 0   50 0
50 1  52 0  52 0   52 0
59 0  67 0  84 1
;

symbol color=black;
proc lifetest data=pilot_certificate plots=(survival);
    time hours*censored(1);
run;
```

The KM estimator of the survival function is given in column Survival. The observations in the column hours that are marked by a star are censored, and the estimates of the survival function remain unchanged at these points. To avoid duplicate values, these estimates are shown as dots. Note that the estimates are the same as computed by hand, except that we rounded the numbers to two decimal places to save space.

hours	Survival
0.0000	1.0000
2.0000*	.
5.0000*	.
8.0000*	.

```
 8.0000*              .
10.0000*              .
16.0000*              .
22.0000*              .
36.0000*              .
40.0000          0.9091
42.0000*              .
47.0000          0.8081
50.0000          0.7071
50.0000*              .
52.0000               .
52.0000               .
52.0000          0.3535
59.0000          0.2357
67.0000          0.1178
84.0000*              .
```

The Kaplan-Meier curve follows. As a default, SAS marks censored observations by a circle.

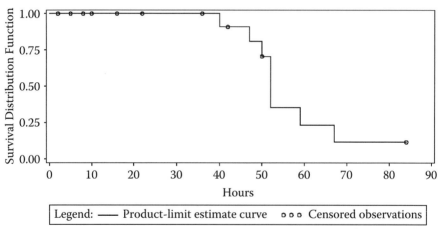

Legend: —— Product-limit estimate curve   o o o Censored observations

From the table and the graph, we can deduce, for instance, that roughly 70% of the students are still in the program after 50 flight hours, and 35%, after 52 hours.   □

6.2 Log-Rank Test for Comparison of Two Survival Functions

In many applications, investigators might want to compare two survival functions. Rigorously speaking, they might want to carry out a two-sided test of statistical hypotheses with $H_0 : S_1(t) = S_2(t)$ for all values of t, and $H_1 : S_1(t) \neq S_2(t)$ for at least one value of t. The *log-rank* test is the most widely used test if some observations are censored. If no censoring is present, the Kolmogorov-Smirnov test is appropriate.

The log-rank test was first proposed in 1966 by an American biostatistician Nathan Mantel (1919-2002).[1]

6.2.1 Testing Procedure

The theoretical background behind the log-rank test is as follows. Let $t_1 < t_2 < \ldots < t_k$ denote the distinct ordered event times in both samples pooled together. At each time t_i, the observations may be summarized in a 2×2 table with the two samples represented by the rows, and the columns representing the number of individuals who experienced the event at time t_i (Event) and the number of those who were at-risk but didn't experience the event (No Event). Thus, there are a total of k 2×2 tables of the form

Sample	Event	No Event	Total
1	e_{1i}	$n_{1i} - e_{1i}$	n_{1i}
2	e_{2i}	$n_{2i} - e_{2i}$	n_{2i}
Total	e_i	$n_i - e_i$	n_i

Testing the null hypothesis of equality of the survival functions is tantamount to simultaneously testing k null hypotheses of independence of the row variable (sample 1 or 2) and the column variable (event or no event) in each table. Under the H_0, e_{1i} has a hypergeometric distribution with parameters n_i (the population size), n_{1i} (the size of the group of interest in the population), and e_i (the sample size). The mean and the variance of e_{1i} are $\mathbb{E}(e_{1i}) = \dfrac{n_{1i}e_i}{n_i}$ and $\mathbb{V}ar(e_{1i}) = \dfrac{n_{1i}n_{2i}(n_i - e_i)e_i}{n_i^2(n_i - 1)}$.

The log-rank test statistic is the standardized sum of e_{1i}'s over all tables, that is,

$$z = \frac{\sum_{i=1}^{k} e_{1i} - \mathbb{E}\left(\sum_{i=1}^{k} e_{1i}\right)}{\sqrt{\mathbb{V}ar\left(\sum_{i=1}^{k} e_{1i}\right)}} = \frac{\sum_{i=1}^{k}\left(e_{1i} - \frac{n_{1i}e_i}{n_i}\right)}{\sqrt{\sum_{i=1}^{k} \frac{n_{1i}n_{2i}(n_i - e_i)e_i}{n_i^2(n_i - 1)}}}.$$

Under H_0 this test statistic has approximately a $\mathcal{N}(0,1)$ distribution. Equivalently, z^2 may be chosen as the test statistic. It has approximately a chi-square distribution with one degree of freedom.

6.2.2 SAS Implementation

The log-rank test statistic z^2, P-value, and plot of the KM survival curves stratified by sample can be requested in SAS by using the following syntax:

```
SYMBOL1 COLOR=black VALUE=none LINE=1; /* solid line*/
SYMBOL2 COLOR=black VALUE=none LINE=2; /*dashed line*/
```

---

[1]Mantel, N. (1966) Evaluation of survival data and two new rank order statistics arising in its consideration, *Cancer Chemotherapy Reports*, **50**, 163-170.

```
PROC LIFETEST DATA=data_name plots=(survival);
    TIME time_variable_name * censoring_variable_name(1);
      STRATA sample_variable_name;
RUN;
```

6.2.3   Example

Example 6.2 Researchers are interested in testing a new antibiotic drug to treat strep throat. Fourteen patients enter the trial. Seven of them are randomly chosen to receive the new drug, whereas the other seven receive a placebo, the traditional ten-day course of a standard antibiotic. The patients are seen by a doctor every day until the strep throat symptoms disappear. The observations are censored for those who still haven't experienced relief at the end of the ten days or were lost to follow-up. The data are

Treatment group (Tx)	3	3	4	4	4+	5	5
Control group (Cx)	4	5	5	6	6	6	10+

The test hypotheses are $H_0 : S_{Tx}(t) = S_{Cx}(t)$ for all t and $H_1 : S_{Tx}(t) \neq S_{Cx}(t)$ for some t. For both groups combined, there are four distinct event times: 3, 4, 5, and 6 days. Therefore, to compute the log-rank test statistic, we first populate four 2×2 tables and calculate the relevant quantities for them. In the context of this example, events are reliefs from strep throat. The calculations follow.

For $t_1 = 3$,

Group	Event	No Event	Total
Tx	2	5	7
Cx	0	7	7
Total	2	12	14

$$e_{11} = 2, \ \mathbb{E}(e_{11}) = \frac{(7)(2)}{14} = 1, \text{ and } \mathbb{V}ar(e_{11}) = \frac{(7)(7)(12)(2)}{(14)^2(13)} = \frac{6}{13}.$$

For $t_2 = 4$,

Group	Event	No Event	Total
Tx	2	3	5
Cx	1	6	7
Total	3	9	12

$$e_{12} = 2, \ \mathbb{E}(e_{12}) = \frac{(5)(3)}{12} = \frac{15}{12}, \text{ and } \mathbb{V}ar(e_{12}) = \frac{(5)(7)(9)(3)}{(12)^2(11)} = \frac{105}{176}.$$

For $t_3 = 5$,

Group	Event	No Event	Total
Tx	2	0	2
Cx	2	4	6
Total	4	4	8

$e_{13} = 2$, $\mathbb{E}(e_{13}) = \dfrac{(2)(4)}{8} = 1$, and $\mathbb{V}ar(e_{13}) = \dfrac{(2)(6)(4)(4)}{(8)^2(7)} = \dfrac{3}{7}$.

For $t_4 = 6$,

Group	Event	No Event	Total
Tx	0	0	0
Cx	3	1	4
Total	3	1	4

$e_{14} = 0$, $\mathbb{E}(e_{13}) = \dfrac{(0)(3)}{4} = 0$, and $\mathbb{V}ar(e_{13}) = \dfrac{(0)(4)(1)(3)}{(4)^2(3)} = 0$. This case may be
omitted from further calculations.

The log-rank test statistic is

$$z = \frac{(2-1) + (2 - 15/12) + (2 - 1)}{\sqrt{(6/13) + (105/176) + (3/7)}} = 2.2554.$$

The two-sided P-value is $2\mathbb{P}(Z > 2.2554) = 0.0241 < 0.05$, hence we reject H_0 and
conclude that the two survival curves differ at some points.

The SAS code for this example is as follows:

```
data strep_throat;
    input group $ days censored @@;
datalines;
Tx 3 0 Tx 3 0 Tx 4 0 Tx 4 0 Tx 4 1 Tx 5 0 Tx  5 0
Cx 4 0 Cx 5 0 Cx 5 0 Cx 6 0 Cx 6 0 Cx 6 0 Cx 10 1
;

symbol1 color=black value=none line=1;
symbol2 color=black value=none line=2;

proc lifetest data=strep_throat plots=(survival);
    time days*censored(1);
        strata group;
run;
```

The log-rank test statistic and the P-value are given as

```
                                 Pr >
Test        Chi-Square     DF    Chi-Square
Log-Rank      5.0868        1      0.0241
```

Note that SAS outputs z^2 as the test statistic and uses the chi-square distribution with one degree of freedom to compute the P-value. The test statistic $z^2 = (2.2554)^2 = 5.0868$ is in agreement with what we computed by hand, and the P-values are the same.

For visual display, the two KM survival curves are plotted on the same graph. Note that after three days the survival curve for the treatment group lies clearly underneath that for the control group. This indicates that the patients taking the new drug recover from strep throat faster.

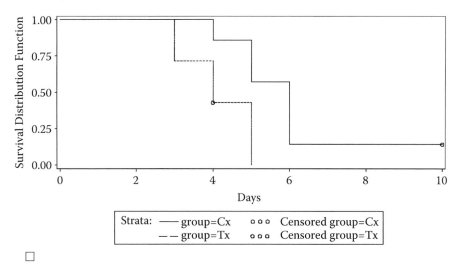

□

6.3   Cox Proportional Hazards Model

Suppose a data set consists of measurements of time to event as well as a number of predictors $x_1(t), \cdots, x_m(t)$. As a rule, some of these predictors depend on time. For instance, in a clinical trial of a cholesterol-lowering medicine, measurements such as patient's gender, race, and age at the time of enrollment do not change with time, whereas blood pressure and vitamin intake are time-dependent variables, varying across the follow-up visits.

If all predictors do not depend on time, the Cox proportional hazards nonparametric model may be fit to the data. This model was founded by a prominent British statistician Sir David Roxbee Cox (1924-) in 1972.[1]

---

[1]Cox, D.R. (1972) Regression models and life-tables, *Journal of the Royal Statistical Society. Series B (Methodological)*, **34**, 187-220.

6.3.1   Two Alternative Definitions of Cox Model

Consider a time to event T with the probability density function $f_T(t)$, cumulative distribution function $F_T(t)$, and survival function $S_T(t) = 1 - F_T(t)$. Define the *hazard function* (or, simply, *hazard*) of T by

$$h_T(t) = \frac{f_T(t)}{S_T(t)}. \tag{6.3}$$

The hazard is interpreted as an *instantaneous event rate* because, given that an individual survives past time t, the probability of the individual experiencing the event within an interval of length dt after time t is approximately equal to the hazard function multiplied by the length of the interval, that is,

$$\mathbb{P}(T < t + dt \mid T > t) = \frac{\mathbb{P}(t < T < t + dt)}{\mathbb{P}(T > t)} \approx \frac{f_T(t)dt}{S_T(t)} = h_T(t)\,dt.$$

Suppose time t and a set of predictors $x_1, \dots, x_m$ not depending on time are observed for an individual. The *Cox proportional hazards model* (or simply *Cox model*) assumes that the hazard function for this individual has the form:

$$h_T(t, x_1, \dots, x_m, \beta_1, \dots, \beta_m) = h_0(t) \exp\{\beta_1 x_1 + \cdots + \beta_m x_m\}. \tag{6.4}$$

Note that in this model the hazard depends on time only through the *baseline hazard function* $h_0(t)$. It represents the hazard function of the *baseline individual*, typically a hypothetical individual with zero values for all predictors.

From (6.4) it is clear that the ratio of hazards of two individuals doesn't depend on time, and thus hazards are *proportional* over time.

The unknowns of this model are the baseline hazard function $h_0(t)$ and regression coefficients $\beta_1, \dots, \beta_m$. Estimating the baseline hazard as a function of time seems like a daunting task. Therefore, we resort to an alternative formulation of the Cox regression model that involves the survival function instead of the hazard function.

Note that $f_T(t) = F_T'(t) = -S_T'(t)$, thus, from (6.3), $h_T(t)$ and $S_T(t)$ are related by a differential equation

$$h_T(t)\,dt = -\frac{S_T'(t)}{S_T(t)}\,dt = -d\ln S_T(t),$$

the solution of which is $S_T(t) = \exp\left\{ -\int_0^t h_T(u)\,du \right\}$. Using this expression and (6.4), we derive the alternative definition of the Cox model,

$$S_T(t, x_1, \dots, x_m, \beta_1, \dots, \beta_m) = \exp\left\{ -\int_0^t h_T(u, x_1, \dots, x_m, \beta_1, \dots, \beta_m)\,du \right\}$$

$$= \exp\left\{ -\int_0^t h_0(u) \exp\{\beta_1 x_1 + \cdots + \beta_m x_m\}\,du \right\} = \left[S_0(t) \right]^r \tag{6.5}$$

where $S_0(t) = \exp\left\{ -\int_0^t h_0(u)\,du \right\}$ is the *baseline survival function*, and $r = \exp\{\beta_1 x_1 + \cdots + \beta_m x_m\}$ is the *relative risk* of an individual. Note that analogous to the baseline hazard function, the baseline survival function corresponds to an individual for whom all predictors have zero values.

6.3.2 Estimation of Regression Coefficients and Baseline Survival Function

The regression coefficients $\beta_1, \ldots, \beta_m$ are estimated by the method of *partial likelihood estimation*, where the portion of the likelihood function that has no time-dependent multiplicative factors is maximized. Suppose distinct ordered event times are $t_1 < \cdots < t_k$. Let $E(t_i)$ denote the set of individuals who experience the event at time t_i, and let $R(t_i)$ be the set of individuals who are at-risk at time t_i. We assume that there are e_i individuals in the set $E(t_i)$. The partial likelihood function has the form:

$$L_p(\beta_1, \ldots, \beta_m) = \prod_{i=1}^{k} \left[\frac{\prod_{j \in E(t_i)} \exp\{\beta_1 x_{1j} + \cdots + \beta_m x_{mj}\}}{\left(\sum_{l \in R(t_i)} \exp\{\beta_1 x_{1l} + \cdots + \beta_m x_{ml}\}\right)^{e_i}} \right].$$

The estimates of the regression coefficients $\widehat{\beta}_1, \ldots, \widehat{\beta}_m$ solve the system of *log-partial-likelihood score equations*,

$$\frac{\partial \ln L_p(\widehat{\beta}_1, \ldots, \widehat{\beta}_m)}{\partial \beta_q} = \sum_{i=1}^{k} \sum_{j \in E(t_i)} x_{qj} -$$

$$- \sum_{i=1}^{k} e_i \frac{\sum_{l \in R(t_i)} x_{ql} \exp\{\widehat{\beta}_1 x_{1l} + \cdots + \widehat{\beta}_m x_{ml}\}}{\sum_{l \in R(t_i)} \exp\{\widehat{\beta}_1 x_{1l} + \cdots + \widehat{\beta}_m x_{ml}\}} = 0, \quad q = 1, \ldots, m. \tag{6.6}$$

Once the estimates of the regression coefficients are obtained, they are used in estimation of the baseline survival function by the method of nonparametric maximum likelihood. Let $\pi_i = \mathbb{P}(T > t_i \mid T > t_{i-1})$, $i = 1, \ldots, k$. The contribution to the likelihood function from an individual with the estimated relative risk $\widehat{r} = \exp\{\widehat{\beta}_1 x_1 + \cdots + \widehat{\beta}_m x_m\}$ who experienced the event at time t_i is $1 - \pi_i^{\widehat{r}}$, and from that who was at risk but did not experience the event, it is $\pi_i^{\widehat{r}}$. Thus, the likelihood function has the form

$$L(\pi_1, \ldots, \pi_k, \widehat{\beta}_1, \ldots, \widehat{\beta}_m) = \prod_{i=1}^{k} \prod_{j \in E(t_i)} \left(1 - \pi_i^{\widehat{r}_j}\right) \prod_{j \in R(t_i) \backslash E(t_i)} \pi_i^{\widehat{r}_j}.$$

Setting the partial derivatives of the log-likelihood function equal to zero and performing some algebraic manipulations, we arrive at a system of normal equations that $\widehat{\pi}_1, \ldots, \widehat{\pi}_k$ satisfy:

$$\sum_{j \in E(t_i)} \frac{\widehat{r}_j}{1 - \widehat{\pi}_i^{\widehat{r}_j}} = \sum_{j \in R(t_i)} \widehat{r}_j, \quad i = 1, \ldots, k. \tag{6.7}$$

This system may be solved numerically. The baseline survival function is then estimated by a step function

$$\widehat{S}_0(t) = \prod_{i:t_i \leq t} \widehat{\pi}_i, \ t \geq 0,$$

and from (6.5), the estimator of the survival function $S_T(t)$ is given by

$$\widehat{S}_T(t) = \left[\prod_{i:t_i \leq t} \widehat{\pi}_i \right]^{\exp\left\{ \widehat{\beta}_1 x_1 + \cdots + \widehat{\beta}_m x_m \right\}}, \ t \geq 0.$$

Note that this estimator generalizes the Kaplan-Meier estimator given by (6.1). If there are no predictors involved, then all $\widehat{r}_j$ in (6.7) are equal to one, and the system of normal equations degenerates to give the KM estimator.

6.3.3 Interpretation of Regression Coefficients

Estimated regression coefficients in the Cox model are interpreted differently depending on whether they correspond to continuous or dummy predictors.

Suppose x_1 is a continuous variable. Then $100\left(\exp\{\beta_1\} - 1\right)\%$ signifies the percent change in hazard function for each unit increase in x_1, provided the other predictors stay constant. Indeed,

$$\frac{h_T(t, x_1 + 1, x_2, \ldots, x_m, \beta_1, \ldots, \beta_m) - h_T(t, x_1, x_2, \ldots, x_m, \beta_1, \ldots, \beta_m)}{h_T(t, x_1, x_2, \ldots, x_m, \beta_1, \ldots, \beta_m)}$$

$$= \frac{h_0(t) \exp\{\beta_1(x_1 + 1) + \beta_2 x_2 + \cdots + \beta_m x_m\}}{h_0(t) \exp\{\beta_1 x_1 + \beta_2 x_2 + \cdots + \beta_m x_m\}} - 1 = \exp\{\beta_1\} - 1.$$

Suppose now x_1 is a dummy variable. Then $100\exp\{\beta_1\}\%$ has the meaning of relative percentage in hazard functions for individuals for whom $x_1 = 1$ and individuals for whom $x_1 = 0$, under the condition that all the other predictors are unchanged. To see this, write

$$\frac{h(t, 1, x_2, \ldots, x_m, \beta_1, \ldots, \beta_m)}{h(t, 0, x_2, \ldots, x_m, \beta_1, \ldots, \beta_m)} = \frac{h_0(t) \exp\{\beta_1 + \beta_2 x_2 + \cdots + \beta_m x_m\}}{h_0(t) \exp\{\beta_2 x_2 + \cdots + \beta_m x_m\}} = \exp\{\beta_1\}.$$

6.3.4 SAS Implementation

The PHREG procedure fits the Cox proportional hazards model. The same model differs slightly in appearance depending on whether dummy variables are created manually or in the CLASS statement. The syntax without the CLASS statement is

```
PROC PHREG DATA=data_name OUTEST=betas_estimates_name;
   MODEL=time_variable_name * censoring_variable_name(1)=list of predictors;
      BASELINE OUT=outdata_name survival=survival function_name;
```

RUN;

• The `outest` option creates a data set containing the estimates of the beta coefficients as well as the P-values for testing their equality to zero. The tests are based on the chi-square distribution with one degree of freedom.
• The `BASELINE` statement outputs the step function $\overline{S}(t)$, say, that is related to the estimate of the baseline function $\widehat{S}_0(t)$ by

$$\overline{S}(t) = \left[\widehat{S}_0(t)\right]^{\exp\{\widehat{\beta}_1 \bar{x}_1 + \cdots + \widehat{\beta}_m \bar{x}_m\}}$$

where $\bar{x}_1, \ldots, \bar{x}_m$ are the sample means of the predictors. From here, the estimate of the survival function $S_T(t)$ can be expressed as

$$\widehat{S}_T(t) = \left[\overline{S}(t)\right]^{\exp\{\widehat{\beta}_1(x_1 - \bar{x}_1) + \cdots + \widehat{\beta}_m(x_m - \bar{x}_m)\}}.$$

The meaning of the function $\overline{S}(t)$ is clear: it is the survival function of a usually hypothetical "average" individual for whom the values of all predictors are equal to the sample means.

• The sample means of the predictors $\bar{x}_1, \ldots, \bar{x}_m$ and the values of the estimated survival function $\overline{S}(t)$ can be retrieved by printing the *outdata_name* file.

Alternatively, if the CLASS statement is inserted into the code preceding the MODEL statement, SAS automatically creates dummy variables for the listed categorical predictors. However, the output doesn't contain the means of these dummy variables. Instead they are absorbed by the estimated survival function. That is, if, for instance, $x_1, \ldots, x_d$ are dummy variables and $x_{d+1}, \ldots, x_m$ are continuous predictors, then the fitted model has the form

$$\overline{\overline{S}}(t) = \left[\widehat{S}_0(t)\right]^{\exp\{\widehat{\beta}_{d+1} \bar{x}_{d+1} + \cdots + \widehat{\beta}_m \bar{x}_m\}}$$

where $\bar{x}_{d+1}, \ldots, \bar{x}_m$ are the sample means of the continuous predictors. Whence, the estimate of the survival function $S_T(t)$ can be expressed as

$$\widehat{S}_T(t) = \left[\overline{\overline{S}}(t)\right]^{\exp\{\widehat{\beta}_1 x_1 + \cdots + \widehat{\beta}_d x_d + \widehat{\beta}_{d+1}(x_{d+1} - \bar{x}_{d+1}) + \cdots + \widehat{\beta}_m(x_m - \bar{x}_m)\}},$$

and the relation holds

$$\overline{S}(t) = \left[\overline{\overline{S}}(t)\right]^{\exp\{\widehat{\beta}_1 \bar{x}_1 + \cdots + \widehat{\beta}_d \bar{x}_d\}}.$$

6.3.5 *Example*

Example 6.3 Oncologists in a large urban hospital are concerned with modeling survival of patients with localized melanomas, a type of skin cancer tumors. They would like to predict the survival outcome in Caucasians with ulcerated melanomas from such prognostic factors as age at time of surgery (in years), gender, lesion site (axial (trunk, neck, head) vs. extremities (arms and legs)), and tumor thickness (in mm). Thirty patients initially treated at this hospital were followed for 5 years and lengths of their survival were recorded. Patients censored prior to the end of the study were either alive but lost to follow-up or died of melanoma unrelated causes. Since all the four factors do not change with time, the Cox proportional hazards model may be fit to the data. The following SAS statements produce the estimate of the survival function.

```
data melanoma;
input age gender $ site $ thickness length censored @@;
   male=(gender='M');
     axial=(site='axial');
  datalines;
31 F axial      3.3 5.0 1   33 F axial      4.2 3.4 0
78 M axial      4.1 3.3 0   32 F extremity 6.1 5.0 1
48 F extremity 3.6 1.2 0   70 M extremity 2.6 5.0 1
57 F axial      1.6 5.0 1   74 M extremity 2.4 2.7 1
71 M axial      2.7 5.0 1   59 M axial      4.2 4.6 1
53 M axial      1.3 5.0 1   55 F extremity 2.7 4.8 0
79 F extremity 1.6 5.0 1   47 F axial      2.2 5.0 1
75 F axial      3.2 2.3 0   78 F extremity 2.7 0.6 1
29 F extremity 1.8 4.4 0   76 F axial      3.6 2.0 0
74 F extremity 4.2 3.5 0   71 M axial      2.5 2.1 1
55 F extremity 2.5 5.0 1   44 M axial      1.9 5.0 1
65 F extremity 2.2 3.3 1   40 F axial      2.2 5.0 1
69 F extremity 3.5 3.7 0   61 M axial      4.4 4.4 1
50 M axial      2.1 4.5 0   85 F extremity 2.4 2.8 1
39 M extremity 5.8 5.0 0   69 M axial      3.5 3.9 0
  ;

proc phreg data=melanoma outest=betas;
   model length*censored(1)=age male axial thickness;
     baseline out=outdata survival=Sbar;
run;

proc print data=betas;
run;

proc print data=outdata;
run;
```

The relevant quantities that SAS outputs are

Parameter	Parameter Estimate	Pr > ChiSq
age	0.04712	0.0488
male	-1.20811	0.1178
axial	0.65729	0.3685
thickness	0.61677	0.0271

age	male	axial	thickness	length	Sbar
58.9	0.4	0.53333	3.03667	0.0	1.00000
58.9	0.4	0.53333	3.03667	1.2	0.97781
58.9	0.4	0.53333	3.03667	2.0	0.95563
58.9	0.4	0.53333	3.03667	2.3	0.92928
58.9	0.4	0.53333	3.03667	3.3	0.89605
58.9	0.4	0.53333	3.03667	3.4	0.85862
58.9	0.4	0.53333	3.03667	3.5	0.82063
58.9	0.4	0.53333	3.03667	3.7	0.77463
58.9	0.4	0.53333	3.03667	3.9	0.72431
58.9	0.4	0.53333	3.03667	4.4	0.67251
58.9	0.4	0.53333	3.03667	4.5	0.61725
58.9	0.4	0.53333	3.03667	4.8	0.55887
58.9	0.4	0.53333	3.03667	5.0	0.50185

The estimated survival function is

$$\widehat{S}_T(t) = \left[\overline{S}(t)\right]^{\widehat{r}}$$

where $\widehat{r} = \exp\{0.04712(\texttt{age} - 58.9) - 1.20811(\texttt{male} - 0.4) + 0.65729(\texttt{axial} - 0.53333) + 0.61677(\texttt{thickness} - 3.03667)\}$.

The estimates of the values of the step function $\overline{S}(t)$ are given in column Sbar. Note that in this model, only age and thickness are significant at the 5% level.

The estimated regression coefficients yield the following interpretation in terms of hazard of dying from melanoma:

• age: For a one-year increase in age, there is a $100(\exp\{0.04712\} - 1)\% = 4.82\%$ increase in the hazard.

• male: For male patients, the hazard function is $100\exp\{-1.20811\}\% = 29.88\%$ of that for female patients.

• axial: For patients with axial melanoma, the hazard is $100\exp\{0.65729\}\%$ $= 192.96\%$ of that for patients with melanoma on extremities.

• thickness: For a one millimeter increase in melanoma thickness, the hazard increases by $100(\exp\{0.61677\} - 1)\% = 85.29\%$.

This model may be used for prediction. For instance, the estimated relative risk for a 50-year-old female with a melanoma on the upper arm that is 2 mm thick is computed as $\hat{r} = \exp\{0.04712(50 - 58.9) - 1.20811(0 - 0.4) + 0.65729(0 - 0.53333) + 0.61677(2 - 3.03667)\} = 0.39611$. For this patient, the probability to survive for 5 years is estimated as $\hat{S}(5) = 0.50185^{0.39611} = 0.7610$.

The same model may be fit by using the CLASS statement. The code is

```
proc phreg data=melanoma outest=betas;
  class gender (ref='F') site/param=ref;
      /* to make 'F' a reference as in the above model*/
  model length*censored(1)=age gender site thickness;
      baseline out=outdata survival=Sbarbar;
run;

proc print data=betas;
run;

proc print data=outdata;
run;
```

The estimated regression coefficients stay the same but the fitted step function differs. Note that in place of the means of the dummy variables, SAS outputs just the names of the reference categories.

		Parameter Estimate	Pr > ChiSq
Parameter			
age		0.04712	0.0488
gender	M	-1.20811	0.1178
site	axial	0.65729	0.3685
thickness		0.61677	0.0271

age	thickness	gender	site	length	Sbarbar
58.9	3.03667	F	extremit	0.0	1.00000
58.9	3.03667	F	extremit	1.2	0.97470
58.9	3.03667	F	extremit	2.0	0.94949
58.9	3.03667	F	extremit	2.3	0.91966
58.9	3.03667	F	extremit	3.3	0.88221
58.9	3.03667	F	extremit	3.4	0.84025
58.9	3.03667	F	extremit	3.5	0.79794
58.9	3.03667	F	extremit	3.7	0.74706

58.9	3.03667	F	extremit	3.9	0.69191
58.9	3.03667	F	extremit	4.4	0.63570
58.9	3.03667	F	extremit	4.5	0.57641
58.9	3.03667	F	extremit	4.8	0.51458
58.9	3.03667	F	extremit	5.0	0.45508

In this case, the estimated survival function is written as

$$\widehat{S}_T(t) = \left[\overline{\overline{S}}(t)\right]^{\widehat{r}}$$

where $\widehat{r} = \exp\{0.04712(\texttt{age} - 58.9) - 1.20811\texttt{male} + 0.65729\texttt{axial} + 0.61677(\texttt{thickness} - 3.03667)\}$. It can be checked that the following identity is true:

$$\overline{S}(t) = \left[\overline{\overline{S}}(t)\right]^{\exp\{(-1.20811)(0.4)+(0.65729)(0.53333)\}} = \left[\overline{\overline{S}}(t)\right]^{0.87574}.$$

□

Exercises for Chapter 6

Exercise 6.1 A clinical trial of a new drug is conducted on 10 patients with advanced pancreatic cancer. The survival times (in weeks) are given below, with a plus sign indicating that the individual was still alive at the end of the trial.

$$3, \ 4, \ 4, \ 4+, \ 8, \ 8+, \ 16, \ 22, \ 24+, \ 30+$$

(a) Compute by hand the Kaplan-Meier estimator of the survival function and plot the KM survival curve. What is the estimated chance of survival beyond 10 weeks?
(b) Use SAS to estimate and plot the survival function.

Exercise 6.2 Engineers in a research department of a large industrial company conduct an accelerated life testing of a new product. This testing is done before the product goes to manufacturing to determine an appropriate warranty period. Translated into years, the ordered lifespans of the tested products are

$$1.1, \ 2.6, \ 2.8, \ 3.1, \ 3.4, \ 3.5, \ 3.5, \ 3.6, \ 3.7, \ 3.8, \ 3.8, \ 4.0, \ 4.1 \ 5.6$$

(a) Estimate the survival function using the KM method of estimation. Do it by hand and in SAS.
(b) Plot the KM survival curve using SAS.
(c) Under what warranty period will the company have to replace 50% of the products? Less than 15% of the products?

Exercise 6.3 An actuary is working on modifying an existing insurance policy for a 10-year term life insurance. This type of insurance pays a benefit if an insured dies within 10 years of policy issue. He draws a random sample from the existing data for the pool of the company's policyholders and compares the distributions of claim time for men and women. An observation is censored if the insured survived longer than 10 years and no benefit has been paid. The data are

Men	4.2	5.0	5.3	6.7	8.2	10+	10+	
Women	3.6	6.7	8.2	9.3	10+	10+	10+	10+

(a) Construct the Kaplan-Meier survival curves stratified by gender. Do calculations and plotting by hand. Comment on what you see on the graph.
(b) Conduct the log-rank test on these data, doing calculations by hand. Clearly state the hypotheses and your conclusion.
(c) Repeat parts (a) and (b) in SAS. Verify that the results are the same.

Exercise 6.4 A prolonged cohort study has been conducted by epidemiologists in a certain geographical area to estimate the incidence of lung cancer in smokers and non-smokers. Observations were censored for those who were lost to follow-up for study unrelated causes, or were lung cancer free at the end of the study. The data (in years) are

Smokers	Non-smokers
1.1	3.6
1.6	4.5
2.1	4.6+
2.4	4.8
2.7	5.7
3.6	5.8
4.7	6.7
4.8+	7.8
5.1+	10.5+
	11.3
	12.6+

(a) Using SAS, construct the two survival curves. Describe their relative positions. Does it appear that non-smoking is a preventive measure for lung cancer?
(b) Using SAS, conduct the log-rank test. Present the hypotheses, test statistic, and P-value. Draw your conclusion.

Exercise 6.5 The following table contains the information about US presidents who are no longer alive. Listed are the name, age at inauguration, the year started office, total years in office, lifespan, and whether was assassinated.
(a) Use the Cox proportional hazards model to regress the lifespan on the other variables. Censor those who were assassinated. Which predictors turn out not to be significant?
(b) Reduce the model to only significant predictor(s). Write down the fitted model

explicitly. Interpret the coefficient(s).
(c) Compute the estimated probability of Thomas Jefferson surviving longer than he did. Use the reduced model from part (b).

Name	Age at Inauguration	Year Started Office	Years in Office	Life-span	Assassinated
Washington, George	57.2	1789	7.8	67.8	0
Adams, John	61.3	1797	4.0	90.7	0
Jefferson, Thomas	57.9	1801	8.0	83.2	0
Madison, James	58.0	1809	8.0	85.3	0
Monroe, James	58.8	1817	8.0	73.2	0
Adams, John Quincy	57.6	1825	4.0	80.6	0
Jackson, Andrew	62.0	1829	8.0	78.2	0
Van Buren, Martin	54.2	1837	4.0	79.6	0
Harrison, William	68.1	1841	0.1	68.1	0
Tyler, John	51.0	1841	3.9	71.8	0
Polk, James	49.3	1845	4.0	53.6	0
Taylor, Zachary	64.3	1849	1.3	65.6	0
Fillmore, Millard	50.5	1850	2.7	74.2	0
Pierce, Franklin	48.3	1853	4.0	64.9	0
Buchanan, James	65.9	1857	4.0	77.1	0
Lincoln, Abraham	52.1	1861	4.1	56.2	1
Johnson, Andrew	56.3	1865	3.9	66.6	0
Grant, Ulysses	46.9	1869	8.0	63.2	0
Hayes, Rutherford	54.4	1877	4.0	70.3	0
Garfield, James	49.3	1881	0.5	49.8	1
Arthur, Chester	52.0	1881	3.5	57.1	0
Cleveland, Grover	48.0	1885	8.0	71.3	0
Harrison, Benjamin	55.5	1889	4.0	67.6	0
McKinley, William	54.1	1897	4.5	58.6	1
Roosevelt, Theodore	42.9	1901	7.5	60.2	0
Taft, William	51.5	1909	4.0	72.5	0
Wilson, Woodrow	56.2	1913	8.0	67.1	0
Harding, Warren	55.3	1921	2.4	57.7	0
Coolidge, Calvin	51.1	1923	5.6	60.5	0
Hoover, Herbert	54.6	1929	4.0	90.2	0
Roosevelt, Franklin	51.1	1933	12.1	63.2	0
Truman, Harry	60.9	1945	7.8	88.6	0
Eisenhower, Dwight	62.3	1953	8.0	78.5	0

(To be continued...)

(Continued...)

Name	Age at Inaugu- ration	Year Started Office	Years in Office	Life- span	Assassi- nated
Kennedy, John	43.6	1961	2.8	46.5	1
Johnson, Lyndon	55.2	1963	5.2	64.4	0
Nixon, Richard	56.0	1969	5.5	81.3	0
Ford, Gerald	61.1	1974	2.5	93.5	0
Reagan, Ronald	70.0	1981	8.0	93.3	0

Exercise 6.6 Patients with deteriorating aortic heart valve receive an equine valve implant via a minimally invasive surgery. Below is the record of age at the time of surgery (in years), gender (Male or Female), diameter of implanted valve (19, 21, 23, 25, 27, or 29 mm), New York Heart Association class (NYHA classification: I=no limitations on physical activity, II=moderate exertion, III=mild exertion, IV=bed rest), duration until death or valve-related complication, and whether a patient was censored. All censoring was due to either death of non-valve related causes, dropping out of the trial for unknown reasons, or being alive without complications at the end of the trial.

(a) Fit the Cox proportional hazards model. Create dummy variables by hand. Write down the fitted model explicitly. Which variables are insignificant at a 5% level? Note that even though some of the predictors are insignificant, investigators might still be interested in controlling for their presence.

(b) Interpret the estimated coefficients for the full model fitted in part (a).

(c) What is a chance of a one-year event-free survival for a 64-year-old woman with NYHA class III who underwent the valve replacement surgery and who needed a valve of diameter 21 mm?

(d) Use the CLASS statement to fit the Cox proportional hazards model. Write it out explicitly. Verify that it is the same model as in part (a).

Age	Gender	Diameter	NYHA class	Duration	Censored
27	F	19	III	0.9	0
56	F	19	II	3.3	1
68	M	25	II	0.8	0
57	F	21	I	2.8	0
56	F	19	III	0.9	0
55	F	19	II	2.5	0
69	F	25	II	4.8	1
48	M	29	II	3.6	1
38	M	25	II	2.0	0
52	F	23	II	3.7	1
72	M	25	II	3.2	1
67	M	21	IV	1.5	0

(To be continued...)

(Continued...)

Age	Gender	Diameter	NYHA class	Duration	Censored
73	F	19	III	3.1	0
47	F	21	III	3.1	1
55	F	21	IV	0.1	0
44	F	19	II	1.7	0
66	F	23	I	4.9	1
53	F	19	I	2.1	0
64	M	29	II	4.1	1
66	M	23	II	1.1	0
38	F	21	I	0.8	0
49	F	27	II	3.2	0
45	M	29	II	1.3	0
74	F	23	II	3.3	1
54	M	19	III	1.7	0
73	F	23	I	1.2	0
52	F	21	I	4.2	1
50	M	25	II	2.8	1
49	M	25	I	4.4	0
61	F	19	II	3.9	1
57	M	23	III	1.6	0
56	M	23	I	2.1	1
45	F	23	IV	0.5	0
63	F	21	II	1.1	0
70	F	23	III	1.6	0
46	M	27	I	4.1	1
71	F	23	II	3.8	1
60	F	23	I	1.0	0
68	F	21	III	2.8	1
62	F	21	II	1.7	0

Chapter 7

Univariate Probability Density Estimation

Let X be a continuous random variable with the probability density function f. Suppose a set of observed realizations of X is available. The objective is to estimate the density f from the given data set. A parametric approach would be to make an assumption on the functional form of the density and estimate only the unknown parameters. If no explicit functional form of f is feasible, a nonparametric approach steps in. We will study two methods, a histogram, first introduced by Karl Pearson in 1895,[1] and a kernel estimator attributed to two American statisticians, Murray Rosenblatt (1926-) and Emanuel Parzen (1929-), who independently introduced this technique in 1956 and 1962, respectively.[2,3]

7.1 Histogram

7.1.1 Definition

Denote by $x_1, \ldots, x_n$ a set of n independent observations of random variable X. To construct a *histogram*, we first subdivide a portion of the real line that includes the range of the data into semi-open intervals, called *bins*, of the form $[x_0 + kh, x_0 + (k+1)h), k = 0, 1, \ldots$. Here x_0 is the *point of origin* of the bins, and h is the *bin width*. Next, we plot a vertical bar above each bin with the height computed as the fraction of the data points within the bin divided by the bin width. A collection of all such vertical bars is called a *histogram*. This name is derived from Greek words "istos" (mast of a ship) and "gramma" (something drawn or written), and points to resemblance between masts and bars of a histogram. Note that a histogram is defined in such a way that the total area of all bars is equal to one.

---

[1]Pearson, K. (1895) Contributions to the mathematical theory of evolution. II. Skew variation in homogeneous material, *Philosophical Transactions of the Royal Society A: Mathematical, Physical and Engineering Sciences*, **186**, 343-326.

[2]Rosenblatt, M. (1956) Remarks on some nonparametric estimates of a density function, *The Annals of Mathematical Statistics*, **27**, 832-837.

[3]Parzen, E. (1962) On estimation of a probability density function and mode, *The Annals of Mathematical Statistics*, **33**, 1065-1076.

The *histogram estimator* $\widehat{f_H}$ of the density function f at a point x is defined as the height of the bin that contains x, that is, as a fraction of observations x_i's lying in the same bin as x, divided by h, the width of the bin. Formally,

$$\widehat{f_H}(x) = \frac{1}{nh} \sum_{i=1}^{n} \mathbb{I}\left(x_i \text{ is in the same bin as } x\right)$$

where $\mathbb{I}$ is an indicator function of the specified event. Simply put, the shaded area of all bars on a histogram comprises the area underneath the density curve.

It is customary to describe the shape of the density as seen on a histogram in terms of *symmetry*, *right-skewness* (a long right tail) or *left-skewness* (a long left tail); *unimodality* (a single peak), *bimodality* (two peaks), or *multimodality* (more than two peaks).

Even though a histogram is the most commonly used method of density estimation, it has a substantial drawback because it depends heavily on the choice of the point of origin x_0 and bin width h.

7.1.2 SAS Implementation

The HISTOGRAM statement in the UNIVARIATE procedure plots a histogram. The syntax is

```
PROC UNIVARIATE DATA=data_name;
  HISTOGRAM variable_name / MIDPOINTS=list of midpoints MIDPERCENTS
                            CFILL=color_name;
RUN;
```

- The heights of the histogram bars reflect percentages of data points.
- The option MIDPOINTS= is followed by a list of midpoints of the bins to be used in calculations. The midpoints must be equally spaced. The list may be given as values separated by blanks, or in the form: *min* TO *max* BY *increment*. If midpoints are not specified, SAS uses $(2n)^{1/3}$ or convenient slightly larger number of bins of equal width over the range of the data.
- The option MIDPERCENTS produces a table of bin midpoints and the corresponding percentage of observations in the bin.
- The option CFILL= specifies the color of the bars on the histogram.

7.1.3 Example

Example 7.1 A Snack Shack at an elementary school is open every Thursday after school helping classes to raise money for field trips. A volunteer runs the Snack

Shack for three years and diligently compiles a record of weekly revenues. The following SAS program reads the instream data and plots a histogram with an automatic bin selection.

```
data snack_shack;
   input revenue @@;
datalines;
437.94  387.51  400.48  403.16  350.87  408.43  275.94
470.83  173.96  423.90  173.70  462.77  343.58  425.04
168.63  392.00  368.24  310.25  403.15  177.03  408.19
175.33  320.00  185.09  462.46  197.78  276.34  392.71
435.85  283.82  383.30  188.31  460.30  180.14  473.08
177.94  457.38  185.24  352.75  400.32  371.00  372.95
425.95  358.55  380.65  377.22  375.36  280.11  450.68
410.33  370.11  380.32  343.44  400.26  227.33  440.37
405.25  425.57  333.21  200.34  433.23  293.51  458.92
190.42  358.06  373.27  373.83  182.70  463.49  350.00
400.04  367.26  167.29  460.23  167.22  400.34  180.03
442.55  190.44  463.85  283.61  350.64  197.66  428.97
183.94  413.37  183.18  465.96  420.45  393.85  433.92
183.60  453.68  203.80  418.52  443.48  407.45  413.35
395.71  410.32  272.41  458.21  283.21  450.92  195.69
223.75  412.15  213.03  240.43  287.79  297.32  296.89
;

proc univariate data=snack_shack;
  histogram revenue/midpercents cfill=gray;
run;
```

The summary of the bin percentages and the histogram are below.

Bin Midpoint	Observed Percent
180	19.643
220	4.464
260	3.571
300	8.036
340	8.929
380	15.179
420	24.107
460	16.071

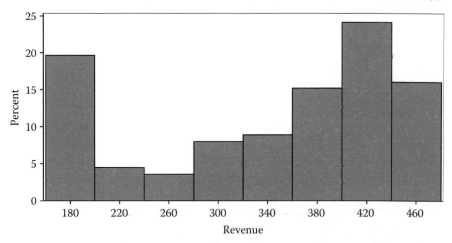

Judging by this histogram, the revenue has a bimodal distribution with one mode around $180 and the other around $420.

If we refine this histogram by specifying more midpoints, we obtain a histogram which suggests that the distribution is rather multimodal than just bimodal. The SAS code and the plot are given below.

```
proc univariate data=snack_shack;
    histogram revenue/midpoints=150 to 450 by 10 cfill=gray;
run;
```

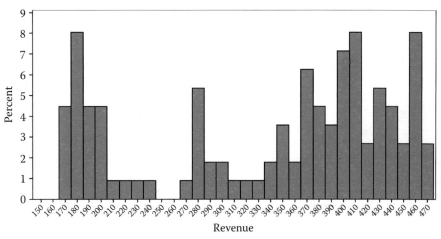

This example illustrates that the choice of the origin and bin width (equivalently, midpoints) is very subjective.   □

7.2 Kernel Density Estimator

7.2.1 Definition

Let $x_1, \ldots, x_n$ be a random sample from a distribution with an unknown density f. A *kernel density estimator* $\widehat{f}_K$ is given by

$$\widehat{f}_K(x) = \frac{1}{n\lambda} \sum_{i=1}^{n} K\left(\frac{x - x_i}{\lambda}\right) \tag{7.1}$$

where λ is called the *bandwidth*, and $K(\cdot)$ is a *smoothing kernel function* which integrates to one, that is, $\int_{-\infty}^{\infty} K(x)\, dx = 1$.

Example 7.2 Suppose a random variable X has a probability density f. At a point x, the function f may be defined as

$$f(x) = \lim_{\lambda \to 0} \frac{1}{2\lambda} \mathbb{P}(x - \lambda < X < x + \lambda).$$

The probability $\mathbb{P}(x - \lambda < X < x + \lambda)$ is naturally estimated by the fraction of the data points within the interval $[x - \lambda, x + \lambda]$. This yields an estimator of the density function at any point x,

$$\widehat{f}_K(x) = \frac{1}{2n\lambda} \sum_{i=1}^{n} \mathbb{I}(x_i \text{ is in the interval } [x - \lambda, x + \lambda]).$$

This estimator may be rewritten in the form (7.1), using a kernel function

$$K(x) = \begin{cases} \frac{1}{2}, & \text{if } |x| \leq 1, \\ 0, & \text{otherwise} \end{cases}.$$

This kernel density estimator is sometimes called a *naive estimator* of f. □

7.2.2 SAS Implementation

SAS computes the kernel estimator of a density function if the KERNEL option in the HISTOGRAM statement is requested. The syntax is

```
PROC UNIVARIATE DATA=data_name;
     VAR variable_name;
HISTOGRAM / <histogram options> KERNEL(C=list of standardized bandwidth val-
ues K=list of kernel names COLOR=list of color names L=list of line types);
RUN;
```

• SAS uses three smoothing kernel functions: *normal*, *quadratic* (or *Epanechnikov* [1]),

---

[1] Epanechnikov, V.A. (1969) Nonparametric estimation of a multidimensional probability density, *Theory of Probability and Its Applications*, **14**, 153-158.

and *triangular*. They are

Normal kernel: $K(x) = \dfrac{1}{\sqrt{2\pi}} \exp\left\{ -\dfrac{1}{2}x^2 \right\}, \quad -\infty < x < \infty.$

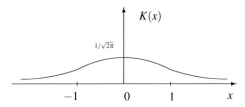

Quadratic (or Epanechnikov) kernel: $K(x) = \begin{cases} \frac{3}{4}(1-x^2), & \text{if } |x| \leq 1, \\ 0, & \text{otherwise} \end{cases}.$

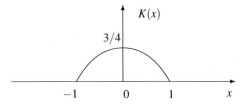

Triangular kernel: $K(x) = \begin{cases} 1 - |x| & \text{if } |x| \leq 1, \\ 0, & \text{otherwise} \end{cases}.$

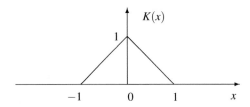

• The option C= lists up to five values of the *standardized bandwidth* c that is related to the bandwidth λ by the formula $\lambda = cQn^{-1/5}$ where Q is the interquartile range of the data. The optimal standardized bandwidth that SAS computes by default is the value of c that minimizes the *approximate mean integrated squared error* (AMISE) defined by

$$AMISE = \frac{(cQ)^4}{4n^{4/5}} \left(\int_{-\infty}^{\infty} x^2 K(x)\,dx \right)^2 \left(\int_{-\infty}^{\infty} \left(g''(x) \right)^2 dx \right) + \frac{1}{cQn^{4/5}} \int_{-\infty}^{\infty} K^2(x)\,dx$$

where g is the normal density with the mean and standard deviation estimated from the sample. The value of the default standardized bandwidth and the corresponding bandwidth value appear in the log window.
• To list the default standardized bandwidth following the option C=, use MISE. For example, C=0.1 0.5 MISE.

• The option K= gives the list of kernels. Three names are possible: normal, quadratic, or triangular. The default value is the normal kernel function.
• If the number of standardized bandwidths specified after C= is larger than the number of kernel functions listed after the option K=, then the last kernel on the list is repeated for the remaining standardized bandwidths. If the number of specified kernels exceeds the number of standardized bandwidths, then the last standardized bandwidth on the list is repeated for the remaining kernels.
• The COLOR= option specifies a list of up to five colors of the density curves.
• The L= option specifies a list of up to five line types of the density curves. The default value is 1 that corresponds to a solid line.

7.2.3   Example

Example 7.3 Returning to Example 7.1, we first construct a histogram and plot the kernel density estimator using all default values. The normal kernel is utilized with an optimal standardized bandwidth $c = 0.7852$ and the corresponding bandwidth $\lambda = 49.831$, as displayed in the log window. The code and the graph are

```
proc univariate data=snack_shack;
    histogram revenue/cfill=gray kernel(color=black);
run;
```

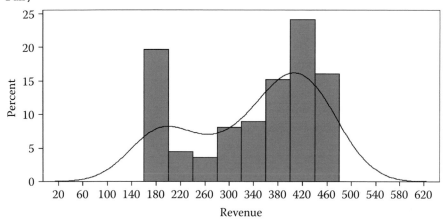

The histogram and the estimated density curve roughly display bimodality. Next, as we did in Example 7.1, to detect multimodality in the data, we input a set of midpoints to increase the number of bins. Then we fit density curves with normal, quadratic, and triangular kernels with standardized bandwidths c=MISE, 0.4, 0.5, and 0.6. From the log window, the values of c that minimize the AMISE are $c = 0.7852$ for the normal kernel, $c = 1.7383$ for the quadratic kernel, and $c = 1.9096$ for the triangular kernel. The code is:

```
title 'c=MISE';
proc univariate data=snack_shack;
```

```
histogram revenue/cfill=gray midpoints= 150 to 500 by 10
   kernel(k=normal quadratic triangular color=black l=1 2 3)
      name='graph1';
run;

title 'c=0.4';
proc univariate data=snack_shack;
histogram revenue/cfill=gray midpoints= 150 to 500 by 10
  kernel(c=0.4 k=normal quadratic triangular color=black
    l=1 2 3) name='graph2';
run;

title 'c=0.5';
proc univariate data=snack_shack;
histogram revenue/cfill=gray midpoints= 150 to 500 by 10
  kernel(c=0.5 k=normal quadratic triangular color=black
    l=1 2 3) name='graph3';
run;

title 'c=0.6';
proc univariate data=snack_shack;
    var revenue;
histogram/ cfill=gray midpoints= 150 to 500 by 10
  kernel(c=0.6 k=normal quadratic triangular color=black
    l=1 2 3) name='graph4';
run;

proc greplay igout=work.gseg tc=sashelp.templt template=l2r2
 nofs;
    treplay 1:graph1 2:graph3 3:graph2 4:graph4;
run;
```

Here are the back-to-back graphs:

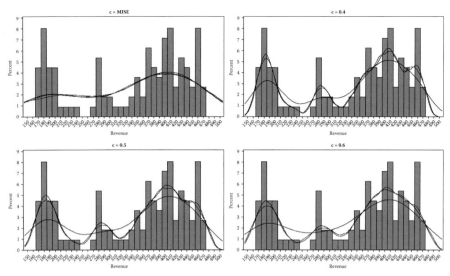

The optimal curves appear to be too smooth. The value of $c = 0.5$ is probably the most satisfactory one. □

Exercises for Chapter 7

Exercise 7.1 An instructor in an introductory statistics course assigned his students a group project which consisted of watching live Old Faithful Geyser at Yellowstone National Park through a webcam and recording times between consecutive eruptions. The times (in minutes) that his students recorded are

65	82	84	54	85	58	79	57	88	68	76	78	74	85	75
65	76	58	83	50	87	78	78	74	66	84	84	98	93	59

(a) Construct a histogram and a normal kernel density curve using the default values for bin width and bandwidth. Comment on the appearance of the graph.

(b) Specify midpoints of a histogram (upon your discretion) and fit a kernel density curve with normal, quadratic, and triangular kernels. Try several standardized bandwidths. Discuss what set of midpoints and what bandwidth give the best results. Is the distribution clearly bimodal?

Exercise 7.2 A professor in the department of Human Development and Family Science studies the distribution of the difference in height between husbands and their wives. She collects the following heights (in inches).

Husband	Wife	Husband	Wife	Husband	Wife
66	63	64	68	77	67
66	65	76	67	64	66
75	62	69	67	72	68
64	68	68	64	68	65
80	71	72	62	76	68
72	69	62	62	65	66
72	66	76	69	68	66
78	70	83	70	77	66
65	61	66	66	77	63
73	70	70	64	75	66
63	63	63	65	73	70
70	68	70	63	76	62
75	69	67	61	64	64
79	66	75	68	77	65
77	64	76	62	63	66
73	70	76	62	74	63
69	68	69	64	80	64
67	70	72	62	64	61
77	71	70	61	69	70
72	62	74	63	79	64

(a) Construct a histogram for the difference in heights. Use the default values in SAS. Describe the density. Does it appear to be symmetric?

(b) Fit a better histogram that discerns right-skewness of the distribution. Model the density using the normal, quadratic, and triangular kernel functions and various bandwidths. Clearly state which model you prefer and why.

Exercise 7.3 Investigators in plant pathology are measuring the length of lesions on rice leaves infected with a fungus *M. grisea*. The lengths in centimeters are

```
2.0 4.1 6.7 4.5 5.7 0.3 6.3 3.5 6.9 2.5 0.9 6.5 6.8 5.4 5.4
3.3 6.9 5.2 2.7 2.9 3.6 2.5 3.8 2.9 3.6 0.9 6.3 3.1 3.7 2.4
1.7 5.4 5.7 4.8 4.9 2.4 3.8 5.3 6.5 6.5 3.2 5.3 4.4 5.4 6.4
0.4 4.2 7.3 7.1 4.4 3.0 2.9 3.4 3.9 3.2 0.6 6.8 6.1 3.0 7.1
3.2 5.6 6.5 7.3 4.2 2.4 7.2 7.1 5.3 5.4 0.3 3.5 6.1 2.8 2.7
0.9 4.0 5.0 3.8 4.1 2.9 6.3 3.4 6.6 4.1 0.3 6.4 4.9 5.9 4.8
3.5 6.0 6.2 7.3 4.2 1.2 4.2 3.5 6.7 3.1 2.4 3.3 4.8 7.1 2.7
1.1 6.6 6.7 3.1 5.8 3.0 5.6 6.6 4.3 6.1 1.9 6.5 7.0 7.0 3.5
2.0 4.5 5.7 5.5 4.6
```

(a) Construct a histogram. Use the default bandwidth. What pattern do you see?

(b) Construct a histogram using an increased number of bins, if appropriate. Comment on the graph. Fit a kernel density estimator with normal, quadratic, and triangular kernels. Which value of the standardized bandwidth c did you select? Explain your choice.

Chapter 8

Resampling Methods for Interval Estimation

Suppose that observations for a random sample are available. The question of interest is to estimate a statistic θ of the underlying distribution. This statistic may be univariate such as mean, variance, median, mode, or bivariate such as correlation coefficient. If the underlying distribution is known, then we can apply developed theories to find a sampling distribution and to give a point estimate and confidence interval in the parametric setting. However, if nothing is known about the distribution, then nonparametric methods must be employed. We consider two relevant techniques: jackknife and bootstrap.

The jackknife method works well when θ is a smooth functional of observations, for instance, mean, variance, or correlation coefficient. If, however, this statistic is non-smooth, such as median or quartile, it is recommended to use the bootstrap method instead. Moreover, the bootstrap is more versatile in the sense that it provides an estimate of the entire sampling distribution of θ, not just an interval estimator.

The name "jackknife" comes simply from the fact that a jackknife is a handy tool that any resourceful person should keep within reach. The origin of the name "bootstrap" is more intricate. The word "bootstrapping" refers to a task completed without external help, such as pulling oneself out of a swamp by the straps of the boots, for instance. Indeed, when the bootstrap method is applied, all the information about the sampling distribution comes from within the sample, without any additional assumptions.

8.1 Jackknife

8.1.1 Estimation Procedure

The *jackknife estimation method* was introduced in 1956 by a British statistician Maurice Henry Quenouille[1](1924-1973) and further extended in 1958 by John

---

[1]Quenouille, M.H. (1956) Notes on bias in estimation, *Biometrika*, **43**, 353-360.

Wilder Tukey (1915-2000), a famous American statistician.[2] The estimation procedure is as follows. Let $\widehat{\theta}$ denote the sample statistic for the original sample $x_1, \ldots, x_n$. For a fixed j, $j = 1, \ldots, n$, introduce the *jackknife replicate* $\widehat{\theta}_j$ as the sample statistic computed for the sample with x_j removed. Define the *pseudovalues* by $\widehat{\theta}_j^* = n\widehat{\theta} - (n-1)\widehat{\theta}_j$, $j = 1, \ldots, n$. The *jackknife estimate* of θ is the average of the pseudovalues, $\widehat{\theta}_J = \dfrac{1}{n} \sum_{i=1}^{n} \widehat{\theta}_j^*$. The variance of the pseudovalues is $\dfrac{\sum_{i=1}^{n} \left(\widehat{\theta}_j^* - \widehat{\theta}_J\right)^2}{(n-1)}$.

An approximate $100(1 - \alpha)\%$ confidence interval for θ based on the t-distribution is computed as

$$\widehat{\theta}_J \pm t_{\alpha/2, n-1} \sqrt{\frac{\sum_{i=1}^{n} \left(\widehat{\theta}_j^* - \widehat{\theta}_J\right)^2}{n(n-1)}}.$$

8.1.2 SAS Implementation

To compute a $100(1 - \alpha)\%$ confidence interval for θ using the jackknife method, we proceed as follows.

Step 1a. If we need to compute a univariate statistic such as mean or median, we use the UNIVARIATE procedure. Here is the syntax:

```
PROC UNIVARIATE DATA=data_name;
    VAR variable_name;
OUTPUT OUT=outdata_name N=n_name statistic_keyword=statistic_name;
RUN;
```

• The option `statistic_keyword` specifies the desired statistic $\widehat{\theta}$. Some examples of keywords are MEAN, MODE, MEDIAN, STD (standard deviation), VAR (variance).

Next, we invoke the CALL SYMPUT routine to create macro variables containing the values of n and $\widehat{\theta}$ as they will be needed later. The syntax is

```
DATA _NULL_;
    SET outdata_name;
        CALL SYMPUT ('n_macro_name', n_name);
            CALL SYMPUT ('statistic_macro_name', statistic_name);
RUN;
```

Step 1b. If the desired statistic θ is a correlation coefficient, then the CORR procedure may be called, for which the syntax is as follows:

---

[2]Tukey, J.W. (1958) Bias and confidence in not-quite large samples (abstract), *The Annals of Mathematical Statistics*, **29**, 614.

```
PROC CORR DATA=data_name OUTP=statistic_data_name;
    VAR variable1_name variable2_name;
RUN;
```

• The option OUTP= specifies that the output data set contain the Pearson correlation coefficient. Alternatively, the option OUTS requests the Spearman correlation coefficient.

The output data set contains several statistics. The syntax below chooses only the sample size n and correlation coefficient $\widehat{\theta}$, and creates macro variables.

```
DATA _NULL_;
    SET statistic_data_name;
IF _TYPE_='N' THEN CALL SYMPUT ('n_macro_name', variable1_name);
IF (_TYPE_='CORR' AND _NAME_='variable1_name')
    THEN CALL SYMPUT ('correlation_macro_name', variable2_name);
RUN;
```

Step 2. We generate the jackknife samples by removing one observation at a time. The syntax is

```
DATA jackknife_samples_name;
    DO sample_count=1 TO &n_macro_name;
        DO record_count=1 TO &n_macro_name;
            SET data_name POINT=record_count;
                IF sample_count NE record_count THEN OUTPUT;
        END;
    END;
        STOP;
RUN;
```

Step 3a. We compute the jackknife replicates $\widehat{\theta}_j$, $j = 1, \ldots, n$. The syntax for the UNIVARIATE procedure is

```
PROC UNIVARIATE DATA=jackknife_samples_name;
    VAR variable_name;
        BY sample_count;
OUTPUT OUT=jackknife_replicates_name statistic_keyword=statistic_name;
RUN;
```

Step 3b. The syntax for the procedure CORR is

```
PROC CORR DATA=jackknife_samples_name; OUTP=jackknife_replicates_name;
    VAR variable1_name variable2_name;
        BY sample_count;
```

```
RUN;
```

Step 4a. We calculate the pseudovalues $\widehat{\theta}_j^*$, $j = 1,\ldots,n$. For the statistic computed by the procedure UNIVARIATE, the syntax is

```
DATA jackknife_replicates_name;
   SET jackknife_replicates_name;
       pseudovalue_name = &n_macro_name * statistic_macro_name
                          - (&n_macro_name - 1) * statistic_name;
   BY sample_count;
   KEEP pseudovalue_name;
RUN;
```

Step 4b. For the correlation coefficient computed by running PROC CORR, use the following syntax.

```
DATA jackknife_replicates_name;
   SET jackknife_replicates_name;
       IF (_TYPE_='CORR' AND _NAME_='variable1_name');
       pseudovalue_name = &n_macro_name * correlation_macro_name
                          -(&n_macro_name - 1) * variable2_name;
   KEEP pseudovalue_name;
RUN;
```

Step 5. We construct a $100(1 - \alpha)\%$ confidence interval for θ based on the t-distribution. The syntax is

```
PROC MEANS DATA=jackknife_replicates_name ALPHA=value;
     VAR pseudovalue_name;
OUTPUT OUT=confidence_interval_name LCLM=lower_limit_name
       UCLM=upper_limit_name;
RUN;

PROC PRINT data=confidence_interval_name;
RUN;
```

• By default, the procedure MEANS computes a 95% confidence interval based on the t-distribution. To change the confidence level, specify a new α in the option ALPHA=.

8.1.3   Examples

Example 8.1 Consider the data on weekly revenues from a Snack Shack considered in Example 7.1. Suppose the objective is to give a 95% confidence interval for the true mean of weekly revenues. As seen on the histograms, the underlying density is multimodal and thus cannot be approximated by some simple parametric curve. We will use the jackknife method to derive the confidence interval. The complete SAS code that uses the data set snack_shack is as follows.

```
proc univariate data=snack_shack;
 var revenue;
   output out=stats n=n_obs mean=mean_all;
run;

data _null_;
  set stats;
    call symput ('n_obs', n_obs);
      call symput ('mean_all', mean_all);
run;

data jackknife_samples;
 do sample=1 to &n_obs;
   do record=1 to &n_obs;
      set snack_shack point=record;
         if sample ne record then output;
   end;
end;
  stop;
run;

proc univariate data=jackknife_samples;
  var revenue;
    by sample;
output out=jackknife_replicates mean=mean_revenue;
run;

data jackknife_replicates;
  set jackknife_replicates;
     pseudomean=&n_obs*&mean_all-(&n_obs-1)*mean_revenue;
  by sample;
    keep pseudomean;
  run;

proc means data=jackknife_replicates;
  var pseudomean;
    output out=CI LCLM=CI_95_lower UCLM=CI_95_upper;
run;

proc print data=CI;
run;
```

The confidence interval is

```
CI_95_      CI_95_
lower       upper
321.578    359.077
   □
```

Example 8.2 For the data in Example 3.3, we apply the jackknife method to construct a 99% confidence interval for the Spearman correlation coefficient between the number of hours students studied for an exam and their scores on that exam. The statements below produce the desired confidence limits.

```
data studying4exam;
input hours score @@;
datalines;
 0 28  5 94 9 84 7 45 17 82 17 99 5 67
12 97 21 79 3 93 7 62 29 60 7 85 10 78
;

proc corr data=studying4exam outs=corr_all;
var hours score;
run;

data _null_;
  set corr_all;
    if (_type_='N') then
    call symput ('n_obs', hours);
    if (_type_='CORR' and _name_='hours') then
    call symput ('corr_all', score);
run;

data jackknife_samples;
 do sample=1 to &n_obs;
   do record=1 to &n_obs;
      set studying4exam point=record;
        if sample ne record then output;
   end;
end;
  stop;
run;

proc corr data=jackknife_samples outs=jackknife_replicates;
   var hours score;
        by sample;
```

```
run;


data jackknife_replicates;
 set jackknife_replicates;
  if(_type_='CORR' and _name_='hours');
     pseudo_corr=&n_obs*&corr_all-(&n_obs-1)*score;
       keep pseudo_corr;
run;

proc means data=jackknife_replicates alpha=0.01;
 var pseudo_corr;
output out=CI lclm=CI_99_lower uclm=CI_99_upper;
run;

proc print data=CI;
run;
```

The confidence interval is

```
CI_99_      CI_99_
 lower       upper
-0.89052    1.20752
```

Note that since a correlation coefficient cannot exceed the value of one, this interval may be truncated from above by one. □

8.2   Bootstrap

8.2.1   Estimation Procedure

The *bootstrap estimation method* was proposed by Bradley Efron (1938-), a famous American statistician, in his seminal article in 1979.[1]

The bootstrap algorithm prescribes to draw *with replacement B* random samples of size n from the original sample $x_1, \ldots, x_n$, thus creating *B bootstrap samples*. For each of these samples, we compute the statistic of interest $\hat{\theta}_b$, $b = 1, \ldots, B$, called the *bootstrap replicate*. It can be shown that the distribution of the bootstrap replicates estimates the sampling distribution of $\hat{\theta}$.

Bootstrapping has a broad range of applications. This procedure may be used, for example, in hypotheses testing, regression analysis, and construction of confidence intervals. Below we describe one method of how bootstrap confidence intervals may be computed.

---

[1]Efron, B. (1979) Bootstrap methods: Another look at the jackknife, *The Annals of Statistics*, **7**, 1-26.

A confidence interval for θ may be constructed directly based on the bootstrap replicates. Denote by $\widehat{\theta}_{L,\alpha/2}$ and $\widehat{\theta}_{U,\alpha/2}$ the $\alpha/2$ and $1 - \alpha/2$ percentiles of the bootstrap replicates, respectively. The $100(1 - \alpha)\%$ *Efron's percentile confidence interval* is

$$\left(\widehat{\theta}_{L,\alpha/2}, \ \widehat{\theta}_{U,\alpha/2}\right).$$

At least 1,000 bootstrap samples are required for a 95% confidence interval, and at least 5,000, for a 99% confidence interval.

8.2.2 *SAS Implementation*

To compute a $100(1 - \alpha)\%$ Efron's percentile confidence interval for θ using the bootstrap method, we proceed as follows.

Step 1. We resort to the SURVEYSELECT procedure to draw bootstrap samples. This procedure is designed to draw random samples by a variety of methods. For bootstrapping, we need the method of sampling with replacement. The syntax is

```
PROC SURVEYSELECT DATA=data_name OUT=bootstrap_samples_name OUTHITS
    SEED=number METHOD=URS SAMPRATE=1 REP=number_of_bootstrap_samples;
RUN;
```

• The output data set *bootstrap_samples_name* contains bootstrap samples indexed by a variable REPLICATE.
• If the option OUTHITS is specified, then each distinct observation in the bootstrap sample appears in the output data set as many times as it was sampled. If this option is omitted, the output data set contains only one record per distinct observation.
• The option SEED specifies a number between 1 and $2^{31} - 2$ to initialize the starting point of the pseudo-random number generator. It is necessary to specify the seed in order to recreate the same bootstrap samples on all subsequent runs of the procedure. If no number is specified, SAS uses the system clock to assign the initial seed, which makes bootstrap samples irreproducible.
• Specifying METHOD=URS requests the *unrestricted random sampling* method which is another name for sampling with replacement.
• The SAMPRATE=1 option assures that drawn samples are all of the same size n as the original sample.
• The option REP= sets the number of bootstrap samples to be drawn.

Step 2a. We compute the bootstrap replicates by using the UNIVARIATE or CORR procedures, whichever is applicable. The syntax for the UNIVARIATE procedure is

```
PROC UNIVARIATE NOPRINT DATA=bootstrap_samples_name;
    VAR variable_name;
      BY REPLICATE;
OUTPUT OUT=bootstrap_replicates_name statistic_keyword=statistic_name;
```

```
RUN;
```

• The option NOPRINT suppresses the output. If it is omitted, then the output for each of the *B* bootstrap samples is printed.
• The statement BY REPLICATE requests to repeat the analysis for each bootstrap sample.

Step 2b. The syntax for the CORR procedure is

```
PROC CORR NOPRINT DATA=bootstrap_samples_name
    OUTP= bootstrap_replicates_name;
        VAR variable1_name variable2_name;
            BY REPLICATE;
RUN;
```

The output data set *bootstrap_replicates_name* contains several statistics. To drop all but the correlation coefficient, we use the following code

```
DATA bootstrap_replicates_name;
    SET bootstrap_replicates_name;
        IF (_TYPE_='CORR' AND _NAME_='variable1_name');
            correlation_name=variable2_name;
        KEEP correlation_name;
RUN;
```

Step 3. We compute a $100(1 - \alpha)\%$ percentile confidence interval of the bootstrap replicates. Here is the syntax.

```
PROC UNIVARIATE DATA=bootstrap_replicates_name;
    VAR statistic_name;
OUTPUT OUT=confidence_interval_name PCTLPRE=list of percentile name prefixes
PCTLPTS=list of percentiles  PCTLNAME=list of percentile name suffixes;
RUN;
```

To view the confidence limits, we print the data set *confidence_interval_name*.

8.2.3   Examples

Example 8.3 We rework Example 8.1, producing a 95% confidence interval for the mean weekly revenues based on the bootstrap estimation method. The SAS code is

```
proc surveyselect data=snack_shack out=bootstrap_samples
   outhits seed=3027574 method=urs samprate=1 rep=1000;
```

```
run;

proc univariate noprint data=bootstrap_samples;
  var revenue;
    by replicate;
      output out=bootstrap_replicates mean=mean_revenue;
run;

proc univariate data=bootstrap_replicates;
  var mean_revenue;
    output out=CI pctlpre=CI_95_ pctlpts=2.5 97.5
            pctlname=lower upper;
run;

proc print data=CI;
run;
```

The output is the 95% Efron's percentile confidence interval.

```
 CI_95_      CI_95_
 lower       upper
321.361     359.832
```

Note that the confidence interval produced by the jackknife method is very similar. □

Example 8.4 Revisiting Example 8.2, we apply the bootstrap method to construct a 99% Efron's percentile confidence interval for the Spearman correlation coefficient between the number of hours students studied for an exam and their scores on that exam. Recall that to produce a reliable interval estimate at the 99% confidence level, it is required to draw a minimum of 5,000 bootstrap samples. The SAS statements are as follows.

```
data studying4exam;
input hours score @@;
datalines;
 0 28  5 94 9 84 7 45 17 82 17 99 5 67
12 97 21 79 3 93 7 62 29 60 7 85 10 78
;

proc surveyselect data=studying4exam out=bootstrap_samples
  outhits seed=234567 method=urs samprate=1 rep=5000;
run;

proc corr noprint data=bootstrap_samples
```

```
  outs=bootstrap_replicates;
    var hours score;
       by replicate;
run;

data bootstrap_replicates;
 set bootstrap_replicates;
  if(_type_='CORR' and _name_='hours');
      spearman_corr=score;
  keep spearman_corr;
run;

proc univariate data=bootstrap_replicates;
  var spearman_corr;
output out=CI pctlpre=CI_99_ pctlpts=0.5 99.5
           pctlname=lower upper;
run;

proc print data=CI;
run;
```

The output is

```
CI_99_         CI_99_
  lower         upper
-0.71292      0.81942
```

Note that the jackknife method gives a wider 99% confidence interval. ☐

Exercises for Chapter 8

Exercise 8.1 For the data in Exercise 7.1, give a 95% confidence interval for the mean time between eruptions of Old Faithful Geyser, using the jackknife estimation procedure.

Exercise 8.2 Use the data in Exercise 7.2 to construct a 99% confidence interval for the Pearson correlation coefficient between heights of husbands and heights of wives. Use the jackknife method of estimation.

Exercise 8.3 Consider the data in Exercise 3.1. Compute the 95% confidence interval for the Spearman correlation coefficient between the prices of gasoline and milk. Apply the jackknife estimation method.

Exercise 8.4 For the data in Exercise 7.3, give the jackknife estimate of the variance of the lengths of lesions and present a 90% confidence interval.

Exercise 8.5 For the data in Exercise 7.1, compute a 95% Efron's percentile confidence interval for the mean time between eruptions of Old Faithful Geyser, using the bootstrap estimation procedure. How does this interval compare to the one computed in Exercise 8.1?

Exercise 8.6 Refer to the data in Exercise 7.2. Construct a 99% Efron's percentile confidence interval for the Pearson correlation coefficient between heights of husbands and heights of their wives. Contrast the result to the jackknife confidence interval from Exercise 8.2.

Exercise 8.7 Consider the data in Exercise 3.1. Give a 95% confidence interval for the Spearman correlation coefficient between the prices of gasoline and milk. Use the method of bootstrap sampling. Is this interval wider or narrower than the one computed in Exercise 8.3 via the jackknife sampling?

Exercise 8.8 Use the data in Exercise 7.3 to produce a 90% confidence interval for the variance of the lengths of lesions based on the bootstrap estimation method. Compare the width of the obtained confidence interval to that of the jackknife interval from Exercise 8.4.

Appendix A

Tables

TABLE A.1 *Critical Values for the Wilcoxon Signed-Rank Test*

	One-Tailed Test		Two-Tailed Test	
n	$\alpha = 0.01$	$\alpha = 0.05$	$\alpha = 0.01$	$\alpha = 0.05$
5	–	0	–	–
6	–	2	–	0
7	0	3	–	2
8	1	5	0	3
9	3	8	1	5
10	5	10	3	8
11	7	13	5	10
12	9	17	7	13
13	12	21	9	17
14	15	25	12	21
15	19	30	15	25
16	23	35	19	29
17	27	41	23	34
18	32	47	27	40
19	37	53	32	46
20	43	60	37	52
21	49	67	42	58
22	55	75	48	65
23	62	83	54	73
24	69	91	61	81
25	76	100	68	89
26	84	110	75	98
27	92	119	83	107
28	101	130	91	116
29	110	140	100	126
30	120	151	109	137

TABLE A.2 *Critical Values for the Wilcoxon Rank-Sum Test*

		One-Tailed Test				Two-Tailed Test			
		$\alpha = 0.01$		$\alpha = 0.05$		$\alpha = 0.01$		$\alpha = 0.05$	
n_1	n_2	W_L	W_U	W_L	W_U	W_L	W_U	W_L	W_U
4	4	–	–	11	25	–	–	10	26
4	5	10	30	12	28	–	–	11	29
4	6	11	33	13	31	10	34	12	32
4	7	11	37	14	34	10	38	13	35
4	8	12	40	15	37	11	41	14	38
4	9	13	43	16	40	11	45	14	42
4	10	13	47	17	43	12	48	15	45
5	5	16	39	19	36	15	40	17	38
5	6	17	43	20	40	16	44	18	42
5	7	18	47	21	44	16	49	20	45
5	8	19	51	23	47	17	53	21	49
5	9	20	55	24	51	18	57	22	53
5	10	21	59	26	54	19	61	23	57
6	6	24	54	28	50	23	55	26	52
6	7	25	59	29	55	24	60	27	57
6	8	27	63	31	59	25	65	29	61
6	9	28	68	33	63	26	70	31	65
6	10	29	73	35	67	27	75	32	70
7	7	34	71	39	66	32	73	36	69
7	8	35	77	41	71	34	78	38	74
7	9	37	82	43	76	35	84	40	79
7	10	39	87	45	81	37	89	42	84
8	8	45	91	51	85	43	93	49	87
8	9	47	97	54	90	45	99	51	93
8	10	49	103	56	96	47	105	53	99
9	9	59	112	66	105	56	115	62	109
9	10	61	119	69	111	58	122	65	115
10	10	74	136	82	128	71	139	78	132

TABLE A.3 *Critical Values for the Ansari-Bradley Test*

		One-Tailed Test				Two-Tailed Test			
		$\alpha = 0.01$		$\alpha = 0.05$		$\alpha = 0.01$		$\alpha = 0.05$	
n_1	n_2	C_L	C_U	C_L	C_U	C_L	C_U	C_L	C_U
4	4	–	–	7	14	–	–	7	14
4	5	7	–	8	15	–	–	7	16
4	6	7	18	8	17	7	18	8	17
4	7	7	20	9	19	7	–	8	19
4	8	7	22	9	20	7	22	8	21
4	9	8	23	10	21	7	24	9	22
4	10	8	25	10	23	8	25	9	24
5	5	10	21	11	20	10	–	11	20
5	6	10	24	12	22	10	24	11	23
5	7	11	25	12	24	10	26	12	24
5	8	11	28	13	26	11	28	12	26
5	9	12	29	14	27	11	30	13	28
5	10	12	32	15	29	11	32	13	30
6	6	14	29	16	27	13	30	15	28
6	7	15	32	17	29	14	32	16	30
6	8	15	34	18	31	15	34	17	32
6	9	16	36	19	34	15	37	17	35
6	10	17	38	19	36	16	39	18	37
7	7	19	38	22	35	18	39	20	37
7	8	20	41	23	38	19	42	21	39
7	9	21	43	24	40	20	44	22	42
7	10	22	46	25	43	21	47	23	44
8	8	25	48	28	45	24	49	27	46
8	9	26	51	30	48	25	52	28	49
8	10	27	54	31	50	26	55	29	52
9	9	32	59	36	55	31	60	34	57
9	10	33	62	37	58	32	64	35	60
10	10	40	71	44	67	39	72	42	69

TABLE A.4 *Critical Values for the Kolmogorov-Smirnov Test*

n_1	n_2	One-Tailed Test		Two-Tailed Test	
		$\alpha = 0.01$	$\alpha = 0.05$	$\alpha = 0.01$	$\alpha = 0.05$
4	4	–	3/4	–	3/4
4	5	4/5	3/4	–	4/5
4	6	5/6	2/3	5/6	3/4
4	7	6/7	5/7	6/7	3/4
4	8	7/8	5/8	7/8	3/4
4	9	7/9	2/3	8/9	3/4
4	10	4/5	13/20	4/5	7/10
5	5	4/5	3/5	4/5	4/5
5	6	5/6	2/3	5/6	2/3
5	7	29/35	23/35	6/7	5/7
5	8	4/5	5/8	4/5	27/40
5	9	7/9	3/5	4/5	31/45
5	10	7/10	3/5	4/5	7/10
6	6	5/6	4/6	5/6	4/6
6	7	5/7	4/7	5/6	29/42
6	8	3/4	7/12	3/4	2/3
6	9	13/18	5/9	7/9	2/3
6	10	7/10	17/30	11/15	19/30
7	7	5/7	4/7	5/7	5/7
7	8	41/56	33/56	3/4	5/8
7	9	5/7	5/9	47/63	40/63
7	10	7/10	39/70	5/7	43/70
8	8	5/8	4/8	6/8	5/8
8	9	2/3	13/24	3/4	5/8
8	10	27/40	21/40	7/10	23/40
9	9	6/9	5/9	6/9	5/9
9	10	2/3	1/2	31/45	26/45
10	10	6/10	5/10	7/10	6/10

TABLE A.5 *Critical Values for the Friedman Rank Test*

k	n	$\alpha = 0.01$	$\alpha = 0.05$	k	n	$\alpha = 0.01$	$\alpha = 0.05$
3	3	–	6.000	4	2	–	6.000
	4	8.000	6.500		3	9.000	7.400
	5	8.400	6.400		4	9.600	7.800
	6	9.000	7.000		5	9.960	7.800
	7	8.857	7.143		6	10.200	7.600
	8	9.000	6.250		7	10.371	7.800
	9	8.667	6.222		8	10.500	7.650
	10	9.600	6.200		9	10.867	7.800
					10	10.800	7.800
5	2	8.000	7.600	6	2	9.714	9.143
	3	10.133	8.533		3	11.762	9.857
	4	11.200	8.800		4	12.714	10.286
	5	11.680	8.960		5	13.229	10.486
	6	11.867	9.067		6	13.619	10.571
	7	12.114	9.143		7	13.857	10.674
	8	12.300	9.300		8	14.000	10.714
	9	12.444	9.244		9	14.143	10.778
	10	12.480	9.280		10	14.229	10.800

TABLE A.6 *Critical Values for the Kruskal-Wallis H-Test*

$k = 3$

n_1	n_2	n_3	$\alpha = 0.01$	$\alpha = 0.05$	n_1	n_2	n_3	$\alpha = 0.01$	$\alpha = 0.05$
4	4	4	7.654	5.692	9	7	5	8.288	5.758
5	4	4	7.760	5.657	9	7	6	8.353	5.783
5	5	4	7.791	5.666	9	7	7	8.403	5.803
5	5	5	8.000	5.780	9	8	4	8.203	5.744
6	4	4	7.795	5.667	9	8	5	8.318	5.783
6	5	4	7.936	5.661	9	8	6	8.409	5.775
6	5	5	8.028	5.699	9	8	7	8.450	5.808
6	6	4	8.000	5.724	9	8	8	8.494	5.810
6	6	5	8.119	5.765	9	9	4	8.202	5.752
6	6	6	8.222	5.719	9	9	5	8.370	5.770
7	4	4	7.814	5.650	9	9	6	8.428	5.814
7	5	4	7.931	5.733	9	9	7	8.469	5.802
7	5	5	8.101	5.708	9	9	8	8.515	5.815
7	6	4	8.016	5.706	9	9	9	8.564	5.845
7	6	5	8.157	5.770	10	4	4	7.907	5.716
7	6	6	8.257	5.730	10	5	4	8.048	5.744
7	7	4	8.142	5.766	10	5	5	8.163	5.777
7	7	5	8.245	5.746	10	6	4	8.143	5.726
7	7	6	8.341	5.793	10	6	5	8.268	5.755
7	7	7	8.378	5.818	10	6	6	8.338	5.780
8	4	4	7.853	5.779	10	7	4	8.172	5.751
8	5	4	7.992	5.718	10	7	5	8.296	5.764
8	5	5	8.116	5.769	10	7	6	8.377	5.799
8	6	4	8.045	5.743	10	7	7	8.419	5.797
8	6	5	8.210	5.750	10	8	4	8.206	5.744
8	6	6	8.294	5.770	10	8	5	8.344	5.789
8	7	4	8.118	5.759	10	8	6	8.398	5.794
8	7	5	8.242	5.777	10	8	7	8.481	5.811
8	7	6	8.333	5.781	10	8	8	8.494	5.829
8	7	7	8.356	5.795	10	9	4	8.223	5.758
8	8	4	8.168	5.743	10	9	5	8.380	5.792
8	8	5	8.297	5.761	10	9	6	8.449	5.813
8	8	6	8.367	5.779	10	9	7	8.507	5.818
8	8	7	8.419	5.791	10	9	8	8.544	5.834
8	8	8	8.465	5.805	10	9	9	8.576	5.831
9	4	4	7.910	5.704	10	10	4	8.263	5.776
9	5	4	8.025	5.713	10	10	5	8.404	5.793
9	5	5	8.170	5.770	10	10	6	8.473	5.796
9	6	4	8.109	5.745	10	10	7	8.537	5.820
9	6	5	8.231	5.762	10	10	8	8.567	5.837
9	6	6	8.307	5.808	10	10	9	8.606	5.837
9	7	4	8.131	5.731	10	10	10	8.640	5.855

$$k = 4$$

n_1	n_2	n_3	n_4	$\alpha = 0.01$	$\alpha = 0.05$
4	4	4	4	9.287	7.235
5	4	4	4	9.392	7.263
5	5	4	4	9.537	7.291
5	5	5	4	9.669	7.328
5	5	5	5	9.800	7.377

TABLE A.7 *Critical Values for the Spearman Rank Correlation Coefficient Test*

	One-Tailed Test		Two-Tailed Test	
n	$\alpha = 0.01$	$\alpha = 0.05$	$\alpha = 0.01$	$\alpha = 0.05$
5	1.000	0.900	–	1.000
6	0.943	0.829	1.000	0.886
7	0.893	0.714	0.929	0.786
8	0.833	0.643	0.881	0.738
9	0.783	0.600	0.833	0.700
10	0.745	0.564	0.794	0.648
11	0.709	0.536	0.755	0.618
12	0.671	0.503	0.727	0.587
13	0.648	0.484	0.703	0.560
14	0.622	0.464	0.675	0.538
15	0.604	0.443	0.654	0.521
16	0.582	0.429	0.635	0.503
17	0.566	0.414	0.615	0.485
18	0.550	0.401	0.600	0.472
19	0.535	0.391	0.584	0.460
20	0.520	0.380	0.570	0.447
21	0.508	0.370	0.556	0.435
22	0.496	0.361	0.544	0.425
23	0.486	0.353	0.532	0.415
24	0.476	0.344	0.521	0.406
25	0.466	0.337	0.511	0.398
26	0.457	0.331	0.501	0.390
27	0.448	0.324	0.491	0.382
28	0.440	0.317	0.483	0.375
29	0.433	0.312	0.475	0.368
30	0.425	0.306	0.467	0.362

Answers to Exercises

Chapter 1

1.1 $M = 7$, P-value $= 0.1719$.
1.2 Valid $n = 7$, $M = 5$, P-value $= 0.2266$.
1.3 $M = 4$, P-value $= 0.375$.
1.4 $T^- = 10$, $T_0 = 10$.
1.5 Valid $n = 7$, $T^- = 4$, $T_0 = 3$.
1.6 $T = 2$, no critical value in Table A.1, do not reject H_0.
1.7 $W = 98$, $W_U = 81$.
1.8 $W = 41$, $W_L = 49$.
1.9 $W = 89$, $W_L = 37$ and $W_U = 89$.
1.10 (a) $W = 81$, $W_L = 49$ and $W_U = 87$; (b) $C = 26$, $C_L = 27$ and $C_U = 46$; (c) $C_L = 28$.
1.11 $W = 69$, $W_L = 42$ and $W_U = 84$; $C = 41$, $C_U = 43$.
1.12 $D^+ = 0.6$ critical value $= 3/5$.
1.13 $D^- = 0.6349$, critical value $= 5/9$.
1.14 $D = 4/10$, critical value $= 43/70$.

Chapter 2

2.1 $Q = 4.7705$, critical value $= 7.8$.
2.2 $Q = 16$, critical value $= 9$, significantly different are Letter and Text, and Phone and Text.
2.3 $H = 11.8552$, critical value $= 5.719$, significantly different are ponds B and C.
2.4 $H = 5.5147$, critical value $= 7.235$.

Chapter 3

3.1 $r_s = 0.9072$, critical value $= 0.671$.
3.2 $r_s = -0.20808$, critical value $= 0.443$.
3.3 $r_s = 0.20909$, critical value $= 0.618$.
3.4 P-value $= 0.0025$.
3.5 P-value $= 0.0027$.

3.6 P-value $= 0.6999$.
3.7 P-value $= 0.6269$.

Chapter 4

4.1 (b) AICC $= -1.13363$, optimal smoothing parameter $= 0.10784$, predicted number of crimes $= 10.40755$; (c) AICC $= -1.09998$, optimal smoothing parameter $= 0.16667$, predicted number of crimes $= 10.40015$.
4.2 (c) Best smoothing parameter $= 0.2$.
4.3 (b) Predicted number of fruits $= 7.85811$.
4.4 Spline with $m = 3$ should be preferred.
4.5 Spline with $m = 2$ should be preferred.
4.6 (b) Predicted number of fruits $= 7.02416$.
4.7 (b) Predicted number of fruits $= 9.76323$.

Chapter 5

5.1 (a) Fitted model is

$$logit(\text{P\_won\_on\_road}) = 0.28378 + 0.05642 * \text{margin\_at\_home}$$
$$+ \text{P\_margin\_at\_home}.$$

(b) Predicted probability $= 0.41994$.
5.2 (a) Fitted model is

$$logit(\text{P\_adherence}) = -4.60560 + 1.05928 * \text{female}$$
$$-5.40983 * \text{no\_intervention} + 0.75363 * \text{age} + \text{P\_age}.$$

(b) Plots are similar for males and females. If there is no intervention applied, the probability of adherence picks around age 9; with intervention, it is the highest between ages 8 and 10.
5.3 (a) Fitted model is

$$logit(\text{P\_PTSD}) = 1.99356 + 2.71395 * \text{female} - 5.73423 * \text{no\_injury}$$
$$+ \text{P\_age\_deployment}.$$

Predicted probability $= 0.90927$.
(b) Fitted model is

$$logit(\text{P\_PTSD}) = -5.37253 + 4.97943 * \text{female} - 8.77597 * \text{no\_injury}$$
$$+ 0.09201 * \text{age} + 0.37085 * \text{deployment} + \text{P\_age} + \text{P\_deployment}.$$

Predicted probability $= 0.99328$.
5.4 (a) Fitted model is

$$\ln \text{P\_n\_accidents} = 0.63437 + 0.30632 * \text{afternoon\_east}$$

$+0.84865 * \texttt{afternoon\_west} + 0.66498 * \texttt{morning\_east} + \texttt{P\_brightness}.$

(b) Fitted model is

$$\ln \texttt{P\_n\_accidents} = 0.35249 + 0.24996 * \texttt{afternoon\_east}$$
$$+ 0.79228 * \texttt{afternoon\_west} + 0.66498 * \texttt{morning\_east}$$
$$+ 0.06060 * \texttt{brightness} + \texttt{P\_brightness}.$$

5.5 (a) Fitted model is

$$\ln \texttt{P\_n\_typos} = 1.85400 + 0.00580 * \texttt{circulation} - 0.28930 * \texttt{cost}$$
$$+ \texttt{P\_circulation} + \texttt{P\_cost}.$$

(b) Predicted number of typos $= 2.31919$.
(d) Fitted model is

$$\ln \texttt{P\_n\_typos} = 1.44734 + \texttt{P\_circulation\_cost}.$$

Predicted number of typos $= 2.44908$.

Chapter 6

6.1 $\widehat{S}(10) = 0.58$.
6.2 (c) The company will have to replace 50% of the products if the warranty period is 3.5 years; less than 15% of the products if the warranty period is 2.6 years (or shorter).
6.3 (a) Most of the survival curve for women lies above that for men, even though they coincide up to 3.6 months, and cross over once at 4.2 months.
(b) Test statistic is $z = 1.0433$, P-value $= 0.2968$.
6.4 (a) The curve for smokers lies clearly underneath that for non-smokers, so non-smokers live longer and, thus, non-smoking is a preventive measure.
(b) Test statistic $= 7.1112$, P-value $= 0.0077$.
6.5 (a) Variables $\texttt{year\_started\_office}$ and $\texttt{year\_in\_office}$ are not significant predictors.
(b) Fitted Cox model has the form

$$\widehat{S}(t) = \left[\overline{S}(t) \right]^{\widehat{r}}$$

where $\widehat{r} = \exp\left\{ -0.13058(\texttt{age\_at\_inauguration} - 55.3895) \right\}$. If the age at inauguration increases by one year, the hazard of dying decreases by 12.24%.
(c) $\widehat{S}(83.2) = 0.20635$.
6.6 (a) Fitted Cox model has the form

$$\widehat{S}(t) = \left[\overline{S}(t) \right]^{\widehat{r}}$$

where $\hat{r} = \exp\{-0.04397(\texttt{age\_at\_surgery} - 56.9) - 0.28846(\texttt{male} - 0.35) - 0.06791(\texttt{valve\_diameter} - 22.6) - 2.81505(\texttt{nyha\_I} - 0.25) - 2.91527(\texttt{nyha\_II} - 0.475) - 2.27741(\texttt{nyha\_III} - 0.2)\}$.

(b) If age increases by one year, hazard decreases by 4.30%; hazard function for males is 74.94% of that for females; if valve diameter were to increase by 1 mm, hazard function would decrease by 6.57%; hazard for NYHA class I is 5.99% of that for class IV; hazard for class II is 5.42% of that for class IV; hazard for class III is 10.25% of that for class IV.

(c) Estimated chance of survival is 84.5%.

(d) Fitted Cox model has the form

$$\widehat{S}_T(t) = \left[\overline{\overline{S}}(t)\right]^{\widehat{r}}$$

where $\hat{r} = \exp\{-0.04397(\texttt{age\_at\_surgery} - 56.9) - 0.28846\,\texttt{male} - 0.06791(\texttt{valve\_diameter} - 22.6) - 2.81505\,\texttt{nyha\_I} - 2.91527\,\texttt{nyha\_II} - 2.27741\,\texttt{nyha\_III}\}$.

Chapter 7

7.1 (a) unimodal, slightly left-skewed, optimal standardized bandwidth equals to 0.7852.

(b) should capture bimodality of data.

7.2 (a) unimodal, roughly symmetric.

7.3 (a) roughly symmetric and possibly bimodal.

Chapter 8

8.1 $[70.14, 79.26]$.

8.2 $[-0.08, 0.56]$.

8.3 $[0.80, 1.00]$.

8.4 $[2.98, 4.16]$.

8.5 close to $[70.55, 78.70]$, depends on chosen seed and number of iterations.

8.6 close to $[-0.09, 0.50]$, depends on chosen seed and number of iterations.

8.7 close to $[0.67, 0.99]$, depends on chosen seed and number of iterations.

8.8 close to $[3.00, 4.13]$, depends on chosen seed and number of iterations.

Recommended Books

Agresti, A. (2012). *Categorical Data Analysis*, John Wiley & Sons, Hoboken, NJ, 3rd edition.

Allison, P.D. (2010). *Survival Analysis Using SAS: A Practical Guide*, SAS Institute, Inc., Cary, NC, 2nd edition.

Corder, G.W. and D.I. Foreman (2009). *Nonparametric Statistics for Non-statisticians: A Step-by-Step Approach*, John Wiley & Sons, Hoboken, NJ.

Efron, B. and R.J. Tibshirani (1994). *An Introduction to Bootstrap*, Chapman & Hall/CRC, Boca Raton, FL.

Higgins, J.J. (2003). *Introduction to Modern Nonparametric Statistics*, Duxbury Press, Florence, KY.

Hollander M., Wolfe D.A., and E. Chicken (2013). *Nonparametric Statistical Methods*, John Wiley & Sons, Hoboken, NJ, 3rd edition.

Richter, S.J. and J.J. Higgins (2005). *SAS Companion for Nonparametric Statistics*, Duxbury Press, Florence, KY.

Sheskin, D.J. (2011). *Handbook of Parametric and Nonparametric Statistical Procedures*, Chapman & Hall/CRC, Boca Raton, FL, 5th edition.

Sprent, P. and N.C. Smeeton (2007). *Applied Nonparametric Statistical Methods*, Chapman & Hall/CRC, Boca Raton, FL, 4th edition.

Index of Notation

Index